Shiva mathe

Applications of Combinatorics

R J Wilson (Editor)

The Open University

Shiva Publishing Limited

69415158
MATH

SHIVA PUBLISHING LIMITED
4 Church Lane, Nantwich, Cheshire CW5 5RQ, England

Hardback edition available in North America from:
BIRKHÄUSER BOSTON, INC.
P.O. Box 2007, Cambridge, MA 02139, USA

British Library Cataloguing in Publication Data

Applications of combinatorics. — (Shiva mathematics series; 6)
  1. Mathematical optimization — Congresses
  2. Combinatorial analysis — Congresses
  I. Wilson, R.J.
  511'.6   QA402.5

ISBN 0–906812–13–5 (paperback)
ISBN 0–906812–14–3 (hardback)

© R.J. Wilson, 1982

Printed in Great Britain by Devon Print Group, Exeter, Devon

Preface

This book presents the proceedings of a one-day conference in
Combinatorics and its Applications held at The Open University,
England, on 13 November 1981. The seven talks presented here were
all given at the conference, and cover a wide variety of topics
ranging from the design of experiments and coding theory to
operational research and chemistry. (In this book the chemistry
talk has been expanded into a longer expository article.) All of
the authors were chosen for their ability to combine interesting
expository material in the areas concerned with an account of
recent research and new results in these areas.

The conference was the first event to be organized by the recently-
formed Open University research group on Combinatorics, and I am
grateful to the members of this group for helping with the arrange-
ments. In particular, I should like to thank Roy Nelson and
Richard Scott for introducing the talks, and to thank the Vice-
Chancellor, Professor J.H. Horlock, F.R.S. for welcoming the par-
ticipants. On the administrative side, I should like to thank
everyone involved, and in particular to Alice Harman and Carole
Fulcher for all their hard work. Most of all, I should like to
express my thanks to Doreen Tucker for her excellent typing of the
entire manuscript, and to Biga Weghofer and the staff of Shiva
Publishing for all their help in producing the final result.

The Open University   Robin J. Wilson
January 1982

Contents

1 Block structures for designed experiments
R.A. BAILEY

Most experiments have their experimental material grouped into
blocks, in one or more ways. In reasonable cases the systems of
blocks form a modular (often distributive) lattice. The lattice
diagram is a useful guide to the analysis of the experiment.

1. BLOCK SYSTEMS

Except in the laboratory, it is a rare experiment in which all the
experimental units are alike. Often the experimenter has to recog-
nize that the units are grouped naturally into blocks of more-or-
less alike units, such as litters when the units are young animals.
In other situations the experimenter imposes "unnatural" blocks in
an attempt to control the variability of a large number of units,
such as plots in a field, where there are no obvious sudden changes
from one to the next. General advice about suitable choice of
blocks may be found in [5], [6] and [8].

There are dangers in ignoring blocks and the variability between
blocks. The variability between blocks is usually different from
that within blocks, and the significance of any treatment differences found has to be assessed in relation to the appropriate type of
underlying variability in the experimental units (the appropriate
type depends on the design used). Unless the types of variability
are differentiated, the significance of apparent treatment differences may be badly under- or over-estimated.

This paper is not concerned with the allocation of treatments to
the experimental units, nor does it deal with the evaluation of
treatment differences. This is because it is important to sort out
the inherent patterns of variability before the treatments are

introduced. Once they are understood, the variability of treatment means may be compared with the appropriate source of inherent variability, using the technique of analysis of variance (see [8], for example).

To make the ideas precise, I make the following definitions.
Definitions. A block system on a (finite) set Ω is a partition of Ω; that is, an equivalence relation on Ω. The equivalence classes are called blocks. The block system is uniform if all its blocks have the same size.

Henceforth, all block systems will be assumed to be uniform.

2. A SINGLE BLOCK SYSTEM

Many experiments have a single block system. This gives the structure known to statisticians as plots-within-blocks.

Example A. In an experiment to compare different diets for young pigs, the experimental units ("plots") are the piglets and the blocks are the litters.

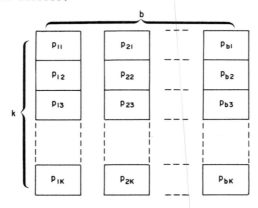

Suppose that there are b blocks and k plots per block. Denote by p_{ij} the j-th plot in the i-th block, for $i = 1, 2, \ldots, b$ and $j = 1, 2, \ldots, k$. My assumption is that plots in the same block (that is, with the same value of i) are more alike than those in different blocks (with different values of i). Thus there are precisely three distinct types of pairs of plots p_{ij} and $p_{i'j'}$:

<u>same</u> plot: $P_{ij} = P_{i'j'}$ $i = i'$, $j = j'$

different plots in the same block $i = i'$, $j \neq j'$

plots in different blocks $i \neq i'$

Note that if $i \neq i'$ then it does not matter whether j is equal to
j' or not as the second subscript signifies only the number of the
plot <u>within</u> a block; there is no connection between, for example,
P_{13} and P_{23}.

In the absence of any real treatment differences the variability
of yields on the plots comes from two sources. The first is the
variability between the block means: this has b - 1 <u>degrees of
freedom</u> (see [8]) (not b, because we deal with the residuals left
when the overall mean is subtracted from all the block means, and
these are constrained to sum to zero). The second source is the
variability of the residuals left when the block means have been
allowed for: since the residuals in each block must sum to zero,
there are k - 1 independent residuals in each block, and so there
are b(k - 1) degrees of freedom for this source. It is often
called the variability of <u>plots-within-blocks</u>.

3. ROWS AND COLUMNS

Another common experimental structure is that of rows and columns.

<u>Example B</u>. Consider an experiment on drugs in which each of n
patients is treated with different drugs in successive months, for
a period of m months altogether. The nm patient-months form the
experimental units, and there are <u>two</u> sorts of block: patients
and months. We can picture the units in a rectangular array, where
the rows represent patients and the columns represent months.

Denote by p_{ij} the experimental unit given by the i-th patient in the j-th month. In contrast with the last example, this time two units which have either suffix in common should be assumed to be more alike than units with both suffices different. Thus there are precisely four types of pairs of unit p_{ij} and $p_{i'j'}$:

same unit	$i = i'$, $j = j'$
different units, same patient	$i = i'$, $j \neq j'$
different units, same month	$i \neq i'$, $j = j'$
different patient, different month	$i \neq i'$, $j \neq j'$

Arguments similar to those used in the last example show that, in the absence of treatment effects, the variability between units can be broken down into three components:

between patients	with n − 1 degrees of freedom
between months	with m − 1 degrees of freedom
between residuals when patients and months have been allowed for	with (n−1)(m−1) degrees of freedom.

4. DOMINANCE

The examples of plots-within-blocks and rows-and-columns may be described in a unified way. In both cases the experimental units are defined by two suffices: the difference is that in the first case (plots-within-blocks) equality of the second suffix has no significance unless the first suffices are also equal. I describe this by saying that suffix 1 dominates suffix 2, or more briefly, 1 dominates 2; this may be pictured thus:

In contrast, in the rows-and-columns structure there is no dominance, so a schematic diagram is

●   ●
I   2

Definition. A subset $\underline{A}$ of the suffices is ancestral if, whenever a suffix J is in $\underline{A}$ and a suffix I dominates J, then I is also in $\underline{A}$. In our two examples the ancestral subsets are:

Plots-within-Blocks	Rows-and-Columns
$\emptyset$	$\emptyset$
{1}	{1}
{1, 2}	{2}
	{1, 2}

The types of pairs of experimental unit correspond to the ancestral subsets. A pair of units has type $\underline{A}$ if all their suffices in $\underline{A}$ are in common, and $\underline{A}$ is the largest ancestral subset with this property. In our two examples the correspondence is as follows:

Plots-within-Blocks		Rows-and-Columns	
$\emptyset$	plots in different blocks	$\emptyset$	different patients, different months
{1}	different plots in the same block	{1}	different units with the same patient
{1, 2}	same plot	{2}	different units with the same month
		{1, 2}	same unit.

The sources of variability also correspond to the ancestral subsets (apart from $\emptyset$, which in a sense corresponds to the single degree of freedom for the overall mean, which is effectively subtracted from all the yields in all calculations of residuals). This correspondence is via the block systems induced by the ancestral subsets.

Definition. Let A be an ancestral subset. The A-block system is the partition into A-blocks, which are maximal sets of experimental units whose suffices in A agree.

For our two examples these types of block are as follows:

Plots-within-Blocks		Rows-and-Columns	
Ø-block	whole set	Ø-block	whole set
{1}-block	block	{1}-block	patient
{1, 2}-block	plot	{2}-block	month
		{1, 2}-block	unit

The plots-within-blocks structure now appears to have _three_ block systems, whereas we previously described it as having only one. This is because we have now included the two trivial block systems in our description: the block systems ε, whose blocks are the singletons, and μ, whose single block is the whole set.

In general, if A is an ancestral subset, the A-source of variability is the variability between means of the A-blocks after allowing for the variability between the means of any types of blocks which contain the A-blocks; that is, the B-blocks for all ancestral subsets B which are properly contained in A. Moreover, the number of degrees of freedom for the A-source of variability is the number of A-blocks minus the numbers of degrees of freedom for all B-sources of variability where B is an ancestral subset properly contained in A (including the single degree of freedom for B = Ø).

For the plots-within-blocks structure these rules give us:

Ancestral subset A	Ancestral subsets B properly contained in A	Number of A-blocks	Source of Variability	Degrees of Freedom
Ø	–	1	overall mean	1
{1}	Ø	b	between blocks	b − 1
{1, 2}	Ø, {1}	bk	between plots within blocks	b(k − 1)

and for the rows-and-columns structure:

Ancestral subset A	Ancestral subsets B properly contained in A	Number of A-blocks	Source of Variability	Degrees of Freedom
$\emptyset$	–	1	overall mean	1
{1}	$\emptyset$	n	between patients	n – 1
{2}	$\emptyset$	m	between months	m – 1
{1, 2}	$\emptyset$, {1}, {2}	nm	between units within patients and months	(n–1)(m–1)

In the analysis of variance there is one term for each source of variability. Because these correspond to the ancestral subsets, in a way that depends upon the inclusions among the ancestral subsets, it is convenient to draw the lattice diagram for the ancestral subsets. If A contains B then each A-block is contained in a B-block, so we place B above A in the lattice: this is the opposite of the usual convention.

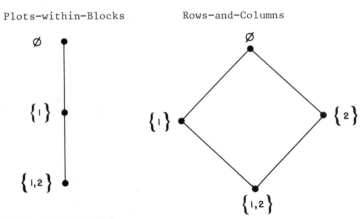

Plots-within-Blocks   Rows-and-Columns

5. PARTIALLY ORDERED SETS

Let S be a set of suffices, some of which may dominate others. From the definition of dominance it is evident that

(i) every suffix in S dominates itself;

(ii) two different suffices in S cannot both dominate each other;

(iii) if I, J, K are suffices and I dominates J and J dominates K, then I dominates K.

Conditions (i)-(iii) tell us that $\underline{S}$ is a <u>partially ordered set</u> (poset).

If $\underline{S} = \{1, 2, \ldots, n\}$ is <u>any</u> finite partially ordered set we can form an array of units $p_{i_1 i_2 \ldots i_n}$, where

$$i_1 = 1, 2, \ldots, m_1$$

$$i_2 = 1, 2, \ldots, m_2$$

$$\cdots \cdots \cdots$$

$$i_n = 1, 2, \ldots, m_n$$

for some integers $m_1, m_2, \ldots, m_n$. With the dominance specified by $\underline{S}$, the rules for finding the different types of pairs of units, the different sources of variability and the degrees of freedom are exactly the rules enunciated in Section 4.

We have already considered all the partially ordered sets with two elements: they give the plots-within-blocks and rows-and-columns structures. We shall now briefly examine the five partially ordered sets with three elements.

Example C

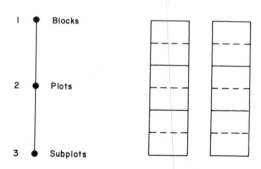

It is not uncommon to perform fertilizer experiments in two stages. At the first stage one type of fertilizer, say different amounts of nitrogen, is applied to the plots. Later each plot is divided into subplots and another type of fertilizer, say different amounts of potash, is applied to the subplots.

Now p_{ijk} denotes the k-th subplot of the j-th plot of the i-th
block. The ancestral subsets are $\emptyset$, $\{1\}$, $\{1, 2\}$, $\{1, 2, 3\}$, with
corresponding block-types the whole set, blocks, plots, subplots,
respectively. The corresponding sources of variability are the
overall mean, between blocks, between plots within blocks, and
between subplots within plots. Differences between the means for
different amounts of nitrogen should be assessed at the plots-
within-blocks level, while potash differences should be assessed
at the subplots-within-plots level.

Example D

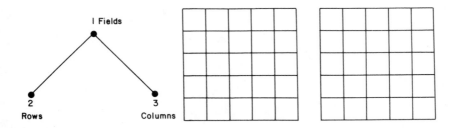

Here p_{ijk} denotes the plot in the j-th row and k-th column of
the i-th field. The ancestral subsets, and their corresponding
sources of variability, are as follows:

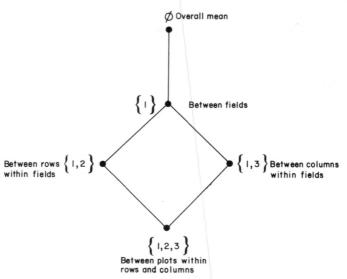

Example E. In a consumer research experiment to compare different brands of vacuum-cleaner, a group of housewives test the vacuum-cleaners over an 8-week period. Each week each housewife uses two of the vacuum-cleaners and compares them. Now an experimental unit is a housewife's use of one vacuum-cleaner for one week, and p_{ijk} denotes the k-th test which the i-th housewife carries out in the

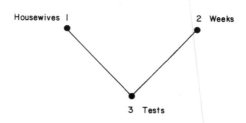

j-th week. The blocks corresponding to the ancestral subset {1, 2} are the housewife-weeks. The complete list of sources of variability is as follows:

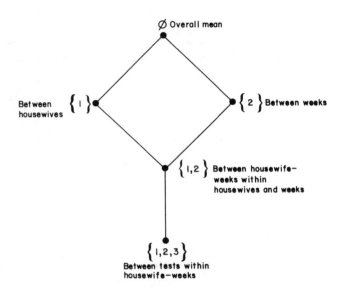

$\emptyset$ Overall mean

Between housewives $\left\{\, 1\, \right\}$

$\left\{\, 2\, \right\}$ Between weeks

$\left\{\, 1,2\, \right\}$ Between housewife—weeks within housewives and weeks

$\left\{\, 1,2,3\, \right\}$ Between tests within housewife—weeks

Example F

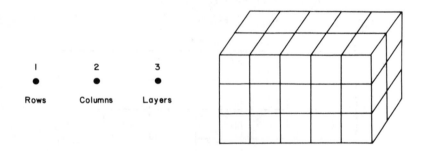

1	2	3
●	●	●
Rows	Columns	Layers

This structure is appropriate for three-dimensional experiments. For example, mushrooms are grown in three-dimensional arrays in dark houses. As there is no dominance, all eight subsets of {1, 2, 3} are ancestral. A selection is shown below, with the corresponding sources of variability. For this purpose I call a {1, 2}-block, which is the intersection of a row with a column, a pillar.

{1}   between rows
{1, 2}   between pillars within rows and columns
{1, 2, 3}   between units within everything else

Example G

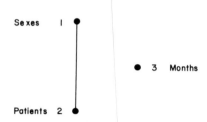

Suppose that the patients in the previously-mentioned drug experiment are of both sexes. Then we may usefully regard the experimental units as a three-way array, with the partially ordered set of suffices shown above. Here p_{ijk} denotes the j-th patient of the i-th sex during the k-th month.

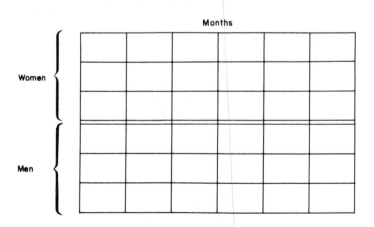

The ancestral subsets and the corresponding sources of variability are shown below.

12

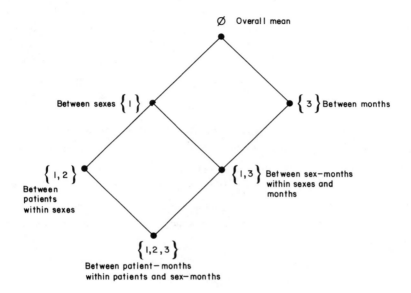

The last example shows the advantage of identifying the sources of variability in a systematic way. I have seen more than one experimenter, dealing with this structure unsystematically, fail to differentiate between the {1, 3}-source of variability and the {1, 2, 3}-source of variability. In the analysis of the subsequent experiments they needed to compare some of the treatment variability with each of these two sources; since they had pooled the estimate of standard deviation from these two sources their results were misleading.

Further details of these structures, and the analyses of experiments performed on them, may be found in [9].

6. BLOCK STRUCTURES

For lattices of more general sets of block systems we need to define the join, ρ ∨ σ, and the meet, ρ ∧ σ, of two block systems ρ and σ.

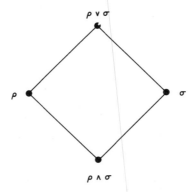

Definitions. Let ρ and σ be two block systems on a set Ω. Then $\rho \wedge \sigma$ is the block system whose blocks are non-empty intersections of ρ-blocks with σ-blocks; and elements α, β of Ω are in the same $\rho \vee \sigma$-block if there are a finite number of elements ω_1, ..., ω_n of Ω such that $\alpha \, \rho \, \omega_1$, $\omega_1 \, \sigma \, \omega_2$, $\omega_2 \, \rho \, \omega_3$, ..., $\omega_n \, \sigma \, \beta$.

Example H below shows that, in general, $\rho \vee \sigma$ is not just the composite relation $\rho\sigma$, for there may be an element ω such that $\alpha \, \rho \, \omega$ and $\omega \, \sigma \, \beta$ but no element ω' such that $\alpha \, \sigma \, \omega'$ and $\omega' \, \rho \, \beta$.

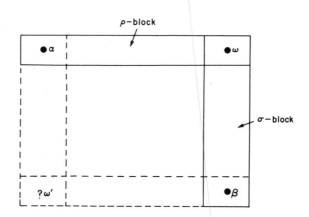

However, if $\rho\sigma = \sigma\rho$ then the existence of such an ω implies the existence of such an ω', and then $\rho \vee \sigma = \rho\sigma$.

Definition. A lattice is <u>modular</u> if, for all ρ, σ, τ with τ above ρ,

$$\tau \wedge (\rho \vee \sigma) = \rho \vee (\tau \wedge \sigma);$$

14

distributive if, for all ρ, σ, τ,

$$\tau \wedge (\rho \vee \sigma) = (\tau \wedge \rho) \vee (\tau \wedge \sigma).$$

Theorem. Every lattice of commuting block systems (ρσ = σρ for all ρ and σ) is modular. (See [7].)

The following example shows that the lattice may not be modular when not all the block systems commute.

Example H. The shaded squares are empty. The ρ-blocks are the rows, the σ-blocks the columns. One τ-block consists of the first and third rows, the other of the second and fourth rows.

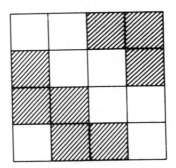

Now ρ ∨ σ = μ, so τ ∧ (ρ ∨ σ) = τ ∧ μ = τ; while τ ∧ σ = ε, so ρ ∨ (τ ∧ σ) = ρ ∨ ε = ρ.

We are concerned chiefly with uniform block systems. The following diagram shows that, even if ρ and σ are both uniform, neither ρ ∧ σ nor ρ ∨ σ (even when ρ ∨ σ = ρσ) need be uniform.

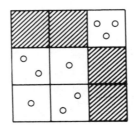

However, if ρ, σ and ρ ∧ σ are uniform, and ρσ = σρ, then ρ ∨ σ is is also uniform (see [11]).

Definitions. A <u>block structure</u> on a set Ω is a set of commuting uniform block systems on Ω which is closed under $\wedge$ and $\vee$. It is a <u>poset block structure</u> if there is a poset $\underline{S}$ and sets Ω_i ($i \varepsilon \underline{S}$) such that $\Omega = \Pi\Omega_i$ and the block systems are given by the ancestral subsets of $\underline{S}$ in the manner described in Sections 4 and 5.

Block structures are discussed more formally in [3]. Poset block structures are discussed in some detail in [11]; they properly contain the <u>simple orthogonal</u> block structures defined in [9] and formalize the structures described in [12] and [13].

Theorem. The lattice of a block structure is distributive if and only if the block structure is a poset block structure. (See [11].)

Although many block structures that occur in statistical experiments are poset block strucures, some other block structures are also common. The simplest is the Latin square structure, which is described in this form in [10].

Example I. If the trees of an experimental orchard have a square layout then the rows and columns form two natural block systems, and the treatments will be applied in a Latin square pattern. However, trees usually show some residual effect of the treatments in the previous year's experiment, which were also applied in a Latin square. The old treatments should therefore be regarded as another block system, and the new treatments applied in a Latin square orthogonal to the one used previously. The lattice of (commuting) block systems is thus:

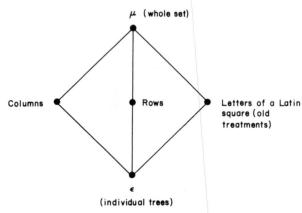

7. AUTOMORPHISMS

To develop a complete statistical theory of the analysis of an experiment with a given block structure it is necessary to know the automorphism group of the structure (see [1], [10]); that is, the set of all permutations of the set Ω which preserve all the block systems in the structure. A poset block structure is determined (up to isomorphism) by the poset $\underline{S}$ and the cardinalities of the sets Ω_i. The automorphism groups of poset block structures are generalized wreath products, which are investigated in [3] and [4]. Automorphism groups of non-poset block structures depend on the individual structure, as the case of non-isomorphic Latin squares of the same size exemplifies. Some results about this case are in [2] and [10].

REFERENCES

1. R.A. Bailey,   A unified approach to design of experiments, J.R. Statist. Soc. A, 144 (1981), 214-223.

2. R.A. Bailey,   Latin squares with highly transitive automorphism groups, J. Austral. Math. Soc. A, 32 (1982), to appear.

3. R.A. Bailey,   Distributive block structures and their automorphisms, Combinatorial Mathematics VIII, (ed. K.L. McAvaney), Lecture Notes in Mathematics 884, Springer-Verlag, Berlin (1981), 115-124.

4. R.A. Bailey, C.E. Praeger, C.A. Rowley and T.P. Speed, Generalized wreath products of permutation groups, to appear.

5. E.G. Cochran and G.B. Cox, Experimental Designs (2nd edition), John Wiley, New York, 1957.

6. D.R. Cox,   Planning of Experiments, John Wiley, New York, 1958.

7. G. Grätzer,   Lattice Theory. First Concepts and Distributive Lattices, W.H. Freeman and Company, San Francisco, 1971.

8. P.W.M. John,   Statistical Design and the Analysis of Experiments, MacMillan, New York, 1961.

9. J.A. Nelder,   The analysis of randomized experiments with orthogonal block structure, Proc. Roy. Soc. A, 283 (1965), 147-178.

10. D.A. Preece, R.A. Bailey and H.D. Patterson, A randomization
        problem in forming designs with superimposed
        treatments, Austral. J. Statist., 20 (1978),
        111 - 125.

11. T.P. Speed and R.A. Bailey, On a class of association schemes
        derived from lattices of equivalence relations,
        Algebraic Structures and Applications (eds.
        P. Schultz, C.E. Praeger and R.P. Sullivan),
        Marcel Dekker, New York (1981), 55-72.

12. T.N. Throckmorton, Structures of Classification Data, Ph.D.
        thesis, Iowa State University, 1961.

13. G. Zyskind, Error Structures in Experimental Designs, Ph.D.
        thesis, Iowa State University, 1958.

R.A. Bailey
Rothamsted Experimental Station
Harpenden
Herts

2 An application of graph colouring on the railways

C. BUTLER and L.R. MATTHEWS

ABSTRACT. An application of graph colouring is described, in the area of the scheduling of work at British Rail Traction Depots. A simple sequential colouring algorithm is used, with minor modifications resulting from the structure of the graph. Although the basic problem reduces to the vertex colouring of a graph and an assignment problem, the numerous different types of practical constraints would preclude the use of such general methods to produce detailed rosters for individual depots: rather the model is useful in investigating the effects of different types of constraints and providing fairly realistic evaluations of various options as part of the long-term review of the organisation of such work into 'links'.

In this article we describe one area in which graph colouring arises in industry. Such applications often involve timetabling or storage problems, and a variety of exact and heuristic algorithms are known ([1]-[3], [5]-[9]). The present work illustrates an area where the objective has not been to produce a general computerised scheduling system. The multiplicity of local conditions at individual depots must shape the actual rosters implemented: however, sufficiently many principles can be abstracted to allow a colouring algorithm to be of use in modelling in broad terms the more general aspects of the organisation of such scheduling.

The drivers at each B.R. Traction Depot work many trains every week, each requiring certain route and traction knowledge, that is, a familiarity with details of the route (location of gradients, signals, etc.) and classes of locomotive or multiple unit to be used. The trains to be run every week are described in terms of diagrams: each diagram may contain several trains and corresponds to the work a driver will carry out in an eight-hour shift. Associated with each diagram are the days of the week on which it runs, the route and traction knowledge required, the signing-on time (time the shift commences), and various other features such as the amount of overtime involved (if any).

A diagram to be run three times a week will involve an eight-hour shift of work on each of three days; these are referred to as turns. Since diagrams (especially where freight trains are concerned) frequently do not run every day, the turns worked by a driver during a week cannot always belong to the same diagram. The work done by an individual driver during a given week is thus a sequence of six turns (excluding Sunday), including one Rest Day, as illustrated by the hypothetical example of Figure 1. Such a sequence is called a line, and the turns are organised into such lines.

Monday	Tuesday	Wednesday	Thursday	Friday	Saturday
Work diagram 503	Work diagram 107	Work diagram 107	Rest Day	Work diagram 122	Work diagram 122

Figure 1

The lines are ordered so that a driver who has completed a line of work will work the next line during the forthcoming week: however, the set of lines is partitioned into subsets called links so that each driver in fact cycles round a subset of lines (his link), rather than working successively through each line in the depot's workload. This is illustrated schematically in Figure 2; such a chart defining lines and links may be modified weekly and daily to produce the actual working plan on any given day.

In practice the links usually contain from around 10 to around 80 lines, although larger links are not uncommon. A depot may have something of the order of 500 turns, divided into from 5 to 10 links.

In addition to the weekly cycling through the lines in a link, a driver may sometimes move from one link to another. For example,

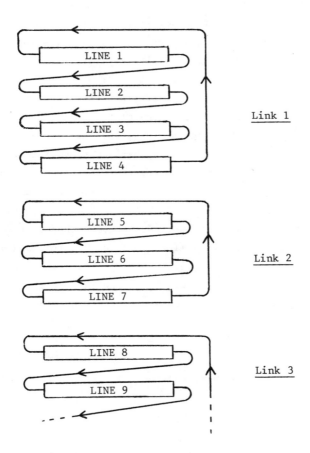

LINE 1
LINE 2
LINE 3
LINE 4

Link 1

LINE 5
LINE 6
LINE 7

Link 2

LINE 8
LINE 9

Link 3

Figure 2

the retirement of a driver from Link 1 may result in a driver moving from Link 2 to Link 1 and a new man, formerly a Driver's Assistant, entering Link 2. Such movements (progression between links) usually involve the learning or re-learning of route and traction knowledge.

There are several reasons for the partition of work into links in this way. A major reason is that each driver need keep up to date only the route and traction knowledge required for this link; however, the links must be at least a certain size in order to avoid monotony and to retain flexibility (for example, in the provision of cover in the event of illness).

It is clear that a given set of lines can be split up into links in many different ways: some of the criteria used are the number and size of the links desired, the distribution of different types of work (and hence earnings) between them, the cost of progression between the links, and the type of spare cover provision (to cover for illness, and also for holidays and trains arranged at short notice).

Consider the turns to be worked. Suppose that there are n_i turns to be worked on day i (for i = 1, 2, ..., 6), and denote these turns by $T_{i,j}$ (j = 1, ..., n_i; i = 1, ..., 6). Let st(i,j) be the signing-on time for turn $T_{i,j}$, and suppose that the turns are numbered in order of increasing signing-on time — that is:

$$st(i,j) \leq st(i,j+1), \text{ for } j = 1, 2, \ldots, n_i-1.$$

We now construct a (simple, undirected) graph on the vertex set

$$\{T_{i,j} \ : \ j = 1, \ldots, n_i; \ i = 1, \ldots, 6\}$$

with edges to be described below; a vertex colouring of this graph will correspond to an assignment of the turns to lines.

Firstly, each line may contain at most one turn for each day of the week. So we have an edge joining $T_{i,j}$ and $T_{i,j'}$, for all j ≠ j' and all i, resulting in six large complete subgraphs.

The remaining edges correspond to constraints on the organisation of work. Some of these arise from legal requirements and national agreements. One of these is the requirement that there should be at least twelve hours between shifts worked by the same man. Since the shifts are eight hours long, this means that the graph has an edge joining $T_{i,j}$ and $T_{i+1,j'}$, whenever

$$st(i,j) - st(i+1,j') > 4 \text{ hours.}$$

Further edges added to the graph correspond to conditions we want to investigate. Often we wish to impose the constraint that vertices $T_{i,j}$ and $T_{i',j'}$ are joined if

$$|st(i,j) - st(i',j')| \geq C,$$

where C is a constant (typically between two and four hours). One aim of the system has been to investigate the effects of different values of such parameters. Such a smoothing out of signing-on times over a week is not always necessary, but is clearly desirable both from the drivers' point of view and for the provision of spare cover.

One of the main variables to be investigated is link structure, which can be specified for the purposes of the algorithm by giving the number of links desired with (upper and lower) bounds on their sizes, and the set of route and traction knowledge required for each one. This then specifies more edges of the graph, each edge joining two turns which cannot belong to the same link and hence cannot belong to the same line. In practice there will be a large degree of overlap in the knowledge required in different links (note that if there were none, the problem would break up into independent problems for each link), and although the assignment of lines to links could in principle reject solutions produced by the colouring algorithm, for practical situations a set of lines (found by colouring the graph) which satisfies the above constraints can readily be assigned to links of the required form.

Finally, other sets of edges are added, corresponding to other constraints at each depot. The number of these varies with the application: only a small number of further types of constraint need be added in order to approximate reasonably well the constraints to be satisfied by a roster in practice: some are already implicit in the link structure imposed.

The colouring algorithm used is a straightforward sequential one, which can be described as follows. The vertices of the graph are first given an ordering v_1, v_2, ..., v_N according to some rule, and the first vertex v_1 is assigned colour 1. The vertices are then considered in turn; each vertex v_r is assigned colour 1 unless one of the previous vertices v_1, ..., v_{r-1} both is adjacent to v_r and has already been assigned colour 1. The same process is now carried out for the uncoloured vertices, starting with the first uncoloured vertex, and assigning colour 2. The operation is repeated using as many colours as necessary until all the vertices have been coloured.

The performance of this method depends on the ordering adopted; for graphs in general, rules for ordering the vertices by decreasing vertex degree, with several alternatives and refinements, have been investigated — a survey is given in [4]. Given the structure of the graphs in the current application the ordering used here was that of increasing signing-on time. This is unaffected by the 'revision' process of [4], although some revision occurs in the tie-breaking described below. With this ordering there is no need to compute vertex degrees explicitly, and adjacency can be checked by referring to signing-on times, lists of available links and a further set of small additional lists.

Each colour is considered to be available to colour at most five vertices: this corresponds to the requirement that each line of work should contain a Rest Day. Two obvious lower bounds for the chromatic number of the graph under this constraint are thus

$$\max_{i=1,2,\ldots,6} \{n_i\} \qquad \text{and} \qquad \left\lceil \sum_{i=1}^{6} \frac{1}{5} n_i \right\rceil,$$

where $\lceil x \rceil$ denotes the least integer greater than or equal to x.
A colouring achieving equality in one of these bounds is clearly
optimal.

Since turns on different days often arise from the same diagram,
ties in the vertex ordering by signing-on time often occur. These
ties are 'broken' by the following heuristic. Suppose that colour
k is the current colour being assigned, and that there is a tie
between several uncoloured vertices with the same signing-on time,
for each of which k is a feasible colour. Denote these competing
vertices by T_{i,m_i}, $i \in I$, where $I \subseteq \{1, 2, \ldots, 6\}$ and $st(i,m_i) = t$
for all $i \in I$. Then, because the distribution throughout the day
of signing-on times is different from day to day (for example,
there are fewer turns running on Saturday afternoons), and because
this affects adjacency, the algorithm attempts to colour vertices
on days that have the most uncoloured days in the 'near future'.
The same constant C is used to define 'near future' as was mentioned in connection with smoothing above. Thus the first vertex
chosen is the vertex T_{i,m_i} which has the largest number of vertices
$(T_{i,j}, j=m_i+1, m_i+2, \ldots)$ satisfying

$$0 < st(i,j) - t \leq C.$$

If this criterion does not resolve the tie then the algorithm simply
colours the turn occurring latest in the week. No further sophis-
tication has been found necessary in practice, nor have other
heuristics investigated proved any more successful.

With the inclusion of the heuristic just outlined, the sequential
process has produced optimal colourings of most graphs found in
practice. With a simpler tie-breaking rule (choosing the latest
available day in the week), one colour more than the optimal number
is generally used.

The colouring algorithm is being used as a small part of an
interactive computer system being developed to estimate the cost of
running a Traction Depot. It should be emphasised that no attempt
is made to produce detailed rosters, nor is this desirable. Apart
from the general considerations of Sunday working, running 'Special'

trains, 'accommodation links' for restricted drivers, and the rostering of Drivers' Assistants and Guards, which have been ignored at present since the constraints on these are very mild, the rosters drawn up by local management take account of local conditions, some (but by no means all) of which are included in the model used. Nevertheless, the chromatic number of the graph gives a measure of the manpower costs and this, together with other considerations such as progression costs and the costs of spare cover, is used to evaluate the broad consequences of various long term (rather than short term) strategies.

These other considerations (progression, spare cover) are regarded as other sub-problems, and solved using different techniques. The colouring algorithm is one of four algorithms, each solving a small part of the overall problem. In several cases (the colouring algorithm being one), the output (solution) from an algorithm is used as input to another program. Using algorithms connected in this way highlights certain qualities that are often desirable for algorithms used in industry but less relevant in theoretical work, and we conclude by listing some of these.

(a) Speed. This requirement is obvious, especially for an interactive system.

(b) Simplicity. This applies both to coding in a high-level programming language and to the form of the algorithm itself, and is important because after the system is developed it must be maintained by people (typically a computing department) who know little of the theoretical basis of the algorithm. Nevertheless, they must be able to assess quickly the effects of any changes that they make to the system.

(c) Robustness. The need to produce 'practical' solutions, where the criteria for practicality change both with time and with use, leads to changes in the properties required of the solutions produced by an algorithm. An algorithm should ideally be robust to such changes in the constraints under which it must operate.

(d) Near-optimality. Because of the difficulty in defining an exact objective function for practical problems, there is

26

frequently little extra value attached to optimal algorithms compared with near-optimal ones. Provided that the solutions can be manipulated (by hand if necessary) to produce a form of solution suitable for implementation, the trade-off between optimality and near-optimality is in terms of inconvenience to the user rather than in terms of better or worse solutions to the problem. (Such manual adjustments are often required for other reasons in any case.) Thus the heuristic tie-breaking rule described is justified in terms of reducing the amount of manual intervention (at the expense of increased computer time) rather than in terms of reducing the costs of running a Traction Depot.

Acknowledgements

The authors would like to thank the British Railways Board for their kind permission to publish this work. We would also like to thank the members of the B.R. Operating Department, in particular Mr. T.A. Greaves and Dr. J. Holt, for introducing us to the problem, and for their help and patience in explaining the practical details.

REFERENCES

1. N. Christofides, An algorithm for the chromatic number of a graph, Computer J. 14 (1971), 38-39.

2. N. Christofides, Graph Theory: an Algorithmic Approach, Academic Press, New York and London, 1976.

3. D.G. Corneil and B. Graham, An algorithm for determining the chromatic number of a graph, SIAM J. Comput. 2 (1973), 311-318.

4. F.D.J. Dunstan, Sequential colourings of graphs, Proc. Fifth British Combinatorial Conference, Congressus Numerantium XV (ed. C. St. J.A. Nash-Williams and J. Sheehan), Utilitas Mathematica, Winnipeg (1976), 151-158.

5. D.W. Matula, G. Marble and J.D. Isaacson, Graph colouring algorithms, Graph Theory and Computing (ed. R.C. Read), Academic Press, New York and London (1972), 109-122.

6. J.E.L. Peck and M.R. Williams, Algorithm 286: Examination Scheduling, Comm. A.C.M. 9 (1966), 433-434.

7. D.J.A. Welsh and M.B. Powell, An upper bound for the chromatic
    number of a graph and its application to
    timetabling problems, Computer J. 10 (1967),
    85-87.

8. M.R. Williams,   The colouring of very large graphs, Combin-
    atorial Structures and their Applications
    (ed. R.K. Guy), Gordon and Breach, New York
    (1970), 477-478.

9. D.C. Wood,   A technique for colouring a graph applicable
    to large scale timetabling problems, Computer
    J. 12 (1969), 317-319.

C. Butler and L. Matthews
Operational Research Division
British Railway Board
11 Tavistock Place
London
WC1

2nd Author's present address is:
Operational Research Department
University of Sussex
Falmer
Brighton

28

3 The travelling salesman problem— a survey

N. CHRISTOFIDES

1. INTRODUCTION

Consider a directed or nondirected graph $G = (N,A)$, where N is a set of vertices and A is a set of arcs. A hamiltonian circuit of G is a circuit passing through every vertex of G exactly once. Given a graph G, the problem of deciding whether it possesses a hamiltonian circuit or not is NP-complete, and no exact algorithm with a polynomial-bounded running time in $n \equiv |N|$ and $m \equiv |A|$ is known for the solution of the problem. If the graph G clearly possesses a hamiltonian circuit (for example, when G is complete), and if the cost of the arcs is given by the matrix $[c_{ij}]$, then the problem of finding the hamiltonian circuit with minimum cost is known as the travelling salesman problem (TSP). The TSP is an archetypal combinatorial optimisation problem with a few real-life applications, but, much more importantly, one which forms the substructure of a large number of other problems arising in practical situations, including many vehicle routing problems [16], [17] and scheduling problems [41].

The purpose of this paper is firstly to review the known results on the existence of hamiltonian circuits in graphs, and secondly to review the exact and approximate procedures that have been suggested for the solution of the TSP. We will concentrate on the algorithmic and computational aspects of these problems, although we will also survey the theoretical aspects of the problem of the existence of hamiltonian circuits. The review of the TSP is based on the unifying viewpoint of relaxation techniques (Lagrangian and state-space) for the derivation of bounds that can be embedded in branch and bound algorithms for solving the problem.

2. THE EXISTENCE OF HAMILTONIAN CIRCUITS IN GRAPHS

2.1 Main direct theorems

The following theorems give some of the known direct results (sufficient conditions) on the existence of hamiltonian circuits in nondirected graphs.

Theorem 1 (Ore [46]). If, for any two non-adjacent vertices a, b of a graph G

$$d(a) + d(b) \geq n,$$

where $d(x)$ is the degree of vertex x, then G is hamiltonian.

Theorem 2 (Erdös-Chvátal [22]). If α and k are, respectively, the independence number and vertex connectivity of a graph G, and if $\alpha \leq k$, then G is hamiltonian.

Theorem 3 (Christofides-Ainouche [19]). Let G be a 2-connected graph, and let a, b, c be any three independent vertices. If

$$\left. \begin{aligned} d(a) + d(b) + d(c) &\geq n + 2 \\ \text{and} \qquad \alpha &\leq \max[3, d_4] \end{aligned} \right\}$$

where d_4 is the fourth smallest degree of G, then G is hamiltonian.

Both Theorems 2 and 3 dominate Theorem 1.

2.2 Closure theorems

Given any condition C, say, a closure of a graph G is the final graph obtained by the recursion

$$G^k = G^{k-1} + (a,b), \quad k = 1, 2, \ldots$$

where a and b are two non-adjacent vertices of G^{k-1} satisfying condition C. (The graph $G^{k-1} + (a,b)$ is the graph obtained by adding arc (a,b) to graph G^{k-1}.) The initial graph $G^0 \equiv G$.

Theorem 4 (Bondy-Chvátal [6]). If the closure of a graph G subject to the condition

$$d(a) + d(b) \geq n$$

is the complete graph, then G is hamiltonian.

Let ℓ_{ab} be the number of vertices of G adjacent to both a and b. Let H_{ab} be the remaining graph after the removal from G of all vertices adjacent to either a or b or both. Let α_{ab} be the independence number of H_{ab}.

Theorem 5 (Ainouche-Christofides [2]). If the closure of a graph G subject to the condition

$$\alpha_{ab} \leq \ell_{ab}$$

is the complete graph, then G is hamiltonian.

Let $\bar{\alpha}_{ab}$ be the number of vertices in H_{ab}.

Theorem 6 (Ainouche-Christofides [2]). If the closure of a graph G subject to the condition

$$\bar{\alpha}_{ab} \leq k$$

is the complete graph, then G is hamiltonian.

Theorem 5 dominates Theorem 4. Stronger versions of Theorems 5 and 6 are found in [1], and a survey of the known results in this area are found in [20].

3. FORMULATIONS OF THE TSP

3.1 Integer programming formulations

Let x_{ij} = 1 if arc (i,j) is in the optimal TSP tour
= 0 otherwise.

The TSP can then be formulated as follows:

31

$$\min \sum_{i=1}^{n} \sum_{j=1}^{n} c_{ij} \, x_{ij} \tag{1}$$

$$\text{s.t.} \sum_{i=1}^{n} x_{ij} = 1, \qquad j \in N \tag{2}$$

$$\sum_{j=1}^{n} x_{ij} = 1, \qquad i \in N \tag{3}$$

$$x_{ij} \in \{0,1\} \qquad i,j \in N \tag{4}$$

$$x_{ij} \text{ must form a tour} \tag{5}$$

This last constraint can be written in a number of different ways, two of which are:

$$\sum_{i \in S_t} \sum_{j \in \bar{S}_t} x_{ij} \geq 1 \qquad \text{for all } S_t \subset N \tag{5a}$$

$$\sum_{i \in S_t} \sum_{j \in \bar{S}_t} x_{ij} \leq |S_t| - 1 \qquad \text{for all } S_t \subset N \tag{5b}$$

where $\bar{S}_t = N - S_t$ and $|S_t|$ is the cardinality of S_t.

Constraints (5a) and (5b) are two expressions of the fact that no subtour through the subset of vertices defined by S_t can exist in any TSP solution. These two sets of constraints are equivalent.

If G is a nondirected graph, let ℓ be the index of the arc set A and let c_ℓ be the corresponding cost. Let $x_\ell = 1$ if arc ℓ is in the solution, and $x_\ell = 0$ otherwise. The symmetric TSP can be formulated as:

$$\min \sum_{\ell=1}^{m} c_\ell x_\ell \tag{6}$$

$$\text{s.t.} \sum_{\ell=1}^{m} x_\ell = n \tag{7}$$

$$\sum_{\ell \varepsilon K_t} x_\ell \geq 2 \quad \text{for all } K_t \equiv (S_t, \overline{S}_t), \quad S_t \subset N \tag{8}$$

$$x_\ell \varepsilon \{0,1\} \qquad \ell = 1, \ldots, m \tag{9}$$

where $K_t \equiv (S_t, \overline{S}_t) = \{(i,j) | i \varepsilon S_t, j \varepsilon \overline{S}_t\}$ is an arc cutset of G.

Constraints (7) and (8) express the fact that on n vertices only a hamiltonian circuit can have n arcs and be 2 arc-connected. The structure of a hamiltonian circuit can also be imposed by alternative equivalent sets of constraints — for example, by the set

$$\sum_{\ell \varepsilon A_i} x_\ell = 2 \qquad i = 1, \ldots, n \tag{10}$$

and $\quad \displaystyle\sum_{\ell \varepsilon K_t} x_\ell \geq 1 \quad \text{for all } K_t = (S, \overline{S}_t), \quad S_t \subset N \tag{11}$

where A_i is the set of arcs incident at vertex i.

Constraints (10) and (11) again express the fact that on n vertices only a hamiltonian circuit has degree 2 for every vertex and is connected.

3.2 Dynamic programming formulation

Let $f(S,i)$ be the cost of a least cost path starting at some specific vertex $v \varepsilon N$ of a graph $G = (N,A)$, passing through every vertex of the set $S \subset N$ exactly once, and finishing at the vertex $i \varepsilon S$.

The TSP can then be formulated as the following dynamic program:

$$f(S,i) = \min_{j \varepsilon S-i} [f(S-i,j) + c_{ji}]. \tag{12a}$$

When the function $f(N,i)$, $i \varepsilon N$, is finally computed from the above recursion, the optimum solution to the TSP can be obtained as:

$$\min_{i \varepsilon N} [f(N,i) + c_{iv}]. \tag{12b}$$

33

Most exact methods for solving the TSP are of the branch and bound variety where, at each node of the branch and bound tree, lower bounds are used in order to limit the search. These bounds are computed by solving related problems which are relaxations of the original TSP (see [36], [37], [3], [11], and [5]). In common with all branch and bound methods, the quality of the computed bounds is the most important parameter determining the efficiency of the resulting algorithm. Various branching schemes for such algorithms are described by Bellmore and Malone [5] and Christofides [12], and will not be discussed here further. Instead we will concentrate our attention on the derivation of bounds using Lagrangian relaxation of the integer programming formulations, and state-space relaxation of the dynamic programming formulations of the TSP.

4. BOUNDS FROM LAGRANGIAN RELAXATIONS

We will describe a number of useful bounds obtained by Lagrangian relaxations of the integer programming formulations of the TSP given in Section 3.1.

4.1 Symmetric TSP — Shortest spanning tree (SST) relaxation

Consider the formulation of the symmetric TSP given by (6), (9), (10) and (11). Let $y_\ell = 1$ if arc ℓ is the longest (greatest cost) arc in the TSP solution, and $y_\ell = 0$ otherwise. Set $x_\ell = 0$ if $y_\ell = 1$. The above-mentioned formulation of the TSP can then be rewritten as:

$$\min \sum_{\ell=1}^{m} c_\ell x_\ell + \sum_{\ell=1}^{m} c_\ell y_\ell \tag{6'}$$

$$\text{s.t.} \sum_{\ell \in A_i} (x_\ell + y_\ell) = 2 \quad i = 1, \ldots, n \tag{10'}$$

$$\sum_{\ell \in K_t} x_\ell \geq 1 \quad \text{for all } K_t = (S_t, \bar{S}_t), \ S_t \subset N \tag{11}$$

$$\sum_{\ell=1}^{m} y_\ell = 1 \tag{13}$$

34

$$x_\ell, \; y_\ell \; \epsilon \; \{0,1\}, \; \ell = 1, \ldots, m \tag{9'}$$

Note that (11) is unchanged, since the solution given by x forms a spanning tree of the graph G.

If $\bar{L}$ is a lower bound on the cost of the longest arc in the TSP solution, then set $y_\ell = 0$ for all ℓ such that $c_\ell < \bar{L}$.

Let λ_i be Lagrange multipliers on constraints (10'). The TSP defined by (6'), (10'), (11), (13) and (9') can then be relaxed to become the two independent problems:

$$P1(\lambda) \begin{cases} \min \; \Sigma(c_\ell + \lambda_{i_\ell} + \lambda_{j_\ell})x_\ell & (14) \\[2ex] \text{s.t.} \; \underset{\ell \epsilon K_t}{\Sigma} \; x_\ell \geq 1 \quad \text{for all } K_t = (S_t, \bar{S}_t), \; S_t \subset N & (15) \end{cases}$$

$$x_\ell \; \epsilon \; \{0,1\} \quad \ell = 1, \ldots, m \tag{16}$$

and

$$P2(\lambda) \begin{cases} \min \; \Sigma(c_\ell + \lambda_{i_\ell} + \lambda_{j_\ell})y_\ell & (17) \\[2ex] \text{s.t.} \; \Sigma y_\ell = 1 & (18) \end{cases}$$

$$y_\ell \; \epsilon \; \{0,1\} \quad \ell = 1, \ldots, m \tag{19}$$

where i_ℓ and j_ℓ are the two terminal vertices of arc ℓ.

Problem $P1(\lambda)$ defined by (14)-(16) is a SST problem on the graph G with the modified costs $c'_\ell = (c_\ell + \lambda_{i_\ell} + \lambda_{j_\ell})$.

Problem $P2(\lambda)$ is solved by simply setting $y_\ell = 1$ for that arc ℓ with the smallest cost $c'_\ell \geq \bar{L}$, where $\bar{L}$ is computed with respect to the modified costs c'_ℓ.

The computation of $\bar{L}$

It is apparent that the longest arc in a TSP solution must be at least as long as the second shortest arc in any cut K_t of G, since

35

every cut K_t must contain at least 2 arcs of the TSP solution. Moreover, consider any vertex $v \in N$. The removal of v from the TSP solution leaves a hamiltonian chain through the remaining $n - 1$ vertices. The longest arc in the TSP solution is, therefore, at least as long as $\max_{X \subset N-v} \{ \min_{\ell \in (S, \bar{S})} [c'_\ell] \}$, where $\bar{S} = N - v - S$, since every cut $(S, \bar{S})$ must contain at least one arc of the TSP solution. Thus, the longest arc in a TSP solution must be at least as large as

$$L^v = \max \left[c'_{\ell_{v2}}, \ \max_{S \subset N-v} \{ \min_{\ell \in (S, \bar{S})} [c'_\ell] \} \right]$$

for any $v \in N$, where $c'_{\ell_{v2}}$ is the cost of the second shortest arc (ℓ_{v2}) incident at vertex v. We can then set $\bar{L} = \max_{v \in N} \{L^v\}$.

Let T be the SST solution to problem $P1(\lambda)$ for a given λ. Let V_1 be the set of vertices of T with degree 1, and $V_2 = N - V_1$.

If $v \in V_1$, set $L^v = c_{\ell_{v2}}$

If $v \in V_2$, compute L^v as $\max_{r=1,..,k} \left[\min_{(T^r, \bar{T}^r)} \{c'_\ell\} \right]$,

where $T^1, \ldots, T^k$ are the subtrees left when vertex v is removed from T. (T^r is used for both the subtree and its set of vertices.) $\bar{L}$ can now be computed as the largest L^v, $v \in N$, and y_ℓ set to 1 for the arc ℓ corresponding to $\bar{L}$.

Since the computation of L^v is expensive for $v \in V_2$, it may be preferable to set $\bar{L} = \max_{v \in V_1} \{L^v\}$.

Computing the bound

For a given λ, let $z(P1(\lambda))$ and $z(P2(\lambda))$ be the optimum solution values of problems $P1(\lambda)$ and $P2(\lambda)$, respectively. A lower bound to the TSP is then

$$z(P_T(\lambda)) = z(P1(\lambda)) + z(P2(\lambda)) - 2 \sum_{i \in N} \lambda_i,$$

where $P_T(\lambda)$ is written for the complete Lagrangian-relaxed problem of the TSP.

Since we are interested in obtaining the highest lower bound possible, we want to choose $\tilde{\lambda}$ which maximises

$$z(P_T(\tilde{\lambda})) = \max z(P_T(\lambda)). \tag{20}$$

Subgradient optimisation (see [52], [38], [35]) and heuristic iterative procedures (see [8], [12]) have been used to approximately maximise (20).

Lagrangian relaxations of the TSP into SST subproblems were first introduced by Christofides [10] and Held and Karp [36], [37] for slightly different versions of the TSP.

4.2 Symmetric TSP — 2-matching relaxation

Once more, consider the formulation of the symmetric TSP given by (6), (9), (10) and (11). Let λ_t be Lagrange multipliers associated with constraints (11), and consider the relaxation of these constraints. The relaxed problem is:

$$\min \sum_{\ell=1}^{m} c_\ell x_\ell + \sum_t \lambda_t (1 - \sum_{\ell \in K_t} x_\ell) \tag{21}$$

subject to (9) and (10).

This relaxed problem is a 2-matching problem which can be solved efficiently by a polynomial algorithm [27], [28].

If initially $\lambda_t = 0$ for all t, then the solution to the relaxed problem will in general be of the form shown in Figure 1. Associated with the 2-matching solution are dual variables u_i on the vertices $i \in N$, and v_b on the blossoms $b \in B$ formed and shrunk during the solution of the 2-matching problem.

From the solution to the relaxed problem one can easily identify constraints of type 11 that are violated and which can, therefore, be relaxed via a Lagrange multiplier λ_t to obtain the modified cost function (23). Figure 1 shows a cut $K_1 = (\{1, \ldots, 4\}, \{5, \ldots, 13\})$ which violates (11). Thus, constraints (11) are formed and relaxed iteratively.

As in the previous relaxation, if $P_M(\lambda)$ is the problem defined by (21), (9) and (10) for a given λ, and if $z(P_m(\lambda))$ is the value of the optimum of $P_M(\lambda)$, we wish to choose $\tilde{\lambda}$ to maximise

$$z(P_M(\tilde{\lambda})) = \max \ [z(P_M(\lambda))].\tag{22}$$

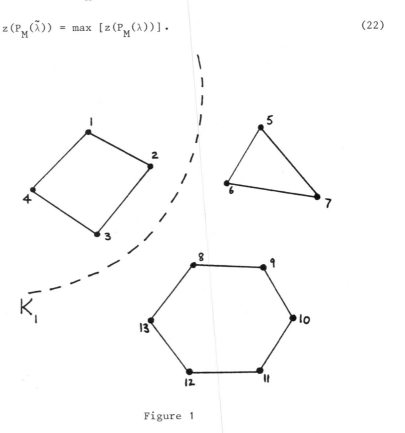

Figure 1

4.3 Restricted Lagrangians

The standard way of performing the maximisation in (20) or (22) is
via subgradient optimisation [49], [38], [35], or variations thereof.
Computationally, this procedure is cumbersome, since it requires
the solution of problems $P_T(\lambda)$ or $P_M(\lambda)$ many times and, for the
violated constraints that are added iteratively, it is difficult
to determine which ones will support a non-zero multiplier. To
alleviate this difficulty, Balas and Christofides [3] introduced
the restricted Lagrangians by imposing on λ the condition that the
solution to problem $P(\lambda)$ is the same as it was for $P(0)$. This
condition has a number of advantages, namely:

38

(a) the problem $P(\lambda)$ has to be solved only once;

(b) the multipliers λ can be computed one at a time, making use of the reduced costs;

(c) the procedure for computing the approximate restricted Lagrangian maximum is polynomial bounded;

(d) alternative optima to $P(\lambda)$ are created which can be used to advantage during the tree search.

Although restricted Lagrangians were introduced in [3] for the asymmetric TSP, they are equally applicable to the symmetric case and, indeed, to any combinatorial problem.

For the case of problem $P_M(\lambda)$, the largest multiplier that can be placed on a violated constraint of type (11) without changing the 2-matching solution is:

$$\bar{\lambda}_t = \min_{\ell \in K_t} \, [\bar{c}_\ell],$$

where $\bar{c}_\ell$ is the reduced cost of arc $\ell \equiv (i_\ell, j_\ell)$ given by

$$\bar{c}_j = c_\ell - u_{i_\ell} - u_j - \sum_{b \in B_\ell} v_b,$$

and where B_ℓ is the set of those blossoms containing both i_ℓ and j_ℓ.

4.4 Asymetric TSP

Corresponding to the SST and 2-matching relaxations of the symmetric TSP are the Shortest Spanning Arborescence (SSA) and Assignment Problem (AP) relaxations of the asymmetric TSP. The correspondence is obvious and we will not dwell on it here further, except to note the main differences:

(a) in general, it is easier (and faster) to compute the SST of a symmetric graph than to compute the SSA of an asymmetric graph of the same size;

(b) in general, it is much easier to solve the AP for an asymmetric graph than to solve the 2-matching for a symmetric one of the same size;

(c) the quality of the bounds obtained (after the Lagrangian ascent) from the SST and AP are better than those from the SSA and

2-matching, respectively.

Thus, in general, tree search algorithms for symmetric problems would make use of the SST bound and those for asymmetric problems would make use of the AP bound.

A detailed analysis of the AP solution to the Lagrangian relaxation problem of asymmetric TSPs has been carried out in [3]. In this reference, the TSP defined by (1)-(5) is relaxed to an AP by relaxing constraints (5). Indeed several versions of (5) are relaxed simultaneously — for example, constraints (5a) are relaxed via Lagrange multipliers λ^a, constraints (5b) are relaxed via Lagrange multipliers λ^b, and other equivalent forms of (5) — for example,

$$\sum_{i \in S_t} \sum_{j \in \bar{S}_t} x_{ij} \geq 1 \quad \text{for all } S_t \subset N-v, \quad \text{for all } v \varepsilon N \qquad (5c)$$
$$\bar{S}_t = N-v-S_t$$

— are relaxed via Lagrange multipliers λ^c. (Note that (5c) imposes the condition that the TSP solution is 1-vertex connected, whereas (5a) imposes 1-arc connectivity.)

Constraints of type (5a) which can support a non-zero multiplier λ^a are then identified one at a time, and the best value of the corresponding λ^a computed. Constraints of type (5b) which can support a non-zero multiplier λ^b are identified, also one at a time, and the best value of λ^b computed; similarly for the constraints of type (5c). In all cases the λ are computed so that the solution to the relaxed problem remains unchanged from that with zero multipliers, as mentioned in Section 4.3. This fast, monotonic, non-iterative ascent in the AP bound (not requiring repeated AP solutions) has made possible the solution of large (in excess of 300 vertices) asymmetric TSPs.

5. BOUNDS FROM STATE SPACE RELAXATIONS

In [18] it is shown that if a mapping function g(.) is used to map the state space (S,i) of a TSP dynamic programming formulation, such as that of (12a), into a lower-dimensional (relaxed) space, then provided $g(.)$ is a separable function, the relaxed recursion provides a lower bound to the value of the TSP.

The mapping functions explored in [18] included, among others:

(a) $g(S,i) = (|S|,i)$.   This is the simplest function where the sets S are simply replaced by their cardinality $k = |S|$.   The DP recursion of (12a) becomes:

$$f(k,i) = \min_{j} \{f(k-1,j) + c_{ji}\} \tag{23}$$

and the bound is

$$LB = \min_{j} \{f(n-1,j) + c_{j1}\}.$$

(b) $g(S,i) = \left(\sum_{j \in S} q_j, i\right)$.

Let arbitrary integers q_j be assigned to the vertices j of the graph G.   This relaxation replaces S by $q = \sum_{j \in S} q_j$.   Clearly if $q_j = 2^j$, this is a one-to-one mapping and is not a relaxation at all.   We are interested only in small positive and negative integers q_j, so that the dimensionality of the relaxed space is small.

The bound is obtained from the recursion

$$f(q,i) = \min_{j} \{f(q-q_i,j) + c_{ji}\} \tag{24}$$

and is given by

$$LB(q) = \min_{j} \{f(\hat{q},j) + c_{j1}\}$$

where $\hat{q} = \sum_{i \in N} q_i$.

5.1   State space ascent

As in Lagrangian relaxation, where the multipliers were changed to obtain the best lower bounds attainable by the method, in state-space relaxation too we can modify the mapping function to maximise the bound. In equation (24), for example, if we insist that $\hat{q}$ remains less than some acceptable upper limit U^q, different choices of q_j will lead to different bounds and we wish to choose $\bar{q}$ which will maximise

$$LB(\bar{q}) = \max_q \{LB(q) \mid \hat{q} \leq U^q\}. \qquad (25)$$

Maximising (25) is more complex than maximising the Lagrangian function, but (25) is more general and has applicability to TSPs with additional constraints.

State-space relaxation is clearly applicable to general combinatorial optimisation problems other than the TSP. The method was developed by Christofides, Mingozzi and Toth [18].

6. LP-BASED ALGORITHMS

Consider the TSP defined by expressions (1), (2), (3), (4) and (5a). One possible procedure for solving the TSP is to solve the linear programming relaxation of TSP (say, $\overline{TSP}$) by dropping constraints (4), and then to impose integrality either in a branch and bound algorithm or by a cutting plane procedure. An obvious problem that immediately arises is due to the very large number of constraints of type (5a) that exists, and which implies that such constraints must be introduced into the LP tableau if and when they are violated by an LP solution rather than in a single step.

The procedure was first suggested by Dantzig et al. [26], and proceeds as follows: relax the TSP as much as possible — for example, by including only constraints (2), (3), and a very small subset of constraints (5a); solve the LP to obtain a solution x. If x represents a hamiltonian circuit the problem is solved; if not, choose a set of inequalities from:

(i) constraints of type (5a) (or (5b)) which are violated by x; and/or

(ii) other constraints which are violated by x — for example, constraints which are satisfied by any integer solution, but which are violated by x, if x turns out to be fractional.

Add the chosen inequalities to the LP tableau and re-optimise
to obtain a new solution x, and repeat until an integer feasible
solution is obtained. The basic method described above has been
adapted in a number of different ways.

Miliotis [44] suggested and tested an algorithm in which
integrality is first achieved — either by branching on fractional
variables, or by introducing Gomory cuts from the constraints in
group (ii) above — and then adding constraints of type (5b) from
group (i).

Grötschel [33] and Crowder and Padberg [24] considered the
simultaneous inclusion of some constraints from (i) and some constraints from (ii). For any (in general, fractional) solution x
to the LP, violated constraints of type (5b) from group (i) were
identified and added to the LP. Constraints from group (ii) were
chosen to be facets of the TSP polytope and, in particular, were
chosen from a class of constraints known as the comb inequalities
(see [21], [34]). These inequalities (which are defined only on
nondirected graphs) are similar to the inequalities introduced by
Edmonds [26] for the linear characterisation of the matching polytope. They eliminate some fractional vertices of the $\overline{TSP}$ polytope,
but may introduce other fractional vertices from their intersections with the constraints of type (5b).

Christofides and Whitlock [15] considered the inclusion, first
of constraints of type (5a) from group (i); only when all constraints of type (5a) were satisfied did they impose constraints
from group (ii) by branching. For a fractional solution x, a
corresponding graph G^x was defined, and by using the Gomory-Hu
algorithm [12] for determining maximum flows between every pair of
vertices of G^x, a number of violated constraints of type (5a) were
identified, or it was shown that all such constraints were satisfied.

In general, LP-based methods are successful, and at least com-
petetive with branch and bound methods, for solving symmetric TSPs.
LP-based methods are not competitive with branch and bound procedures for asymmetric TSPs.

7. HEURISTICS FOR THE TSP

The TSP is an NP-complete problem [31] and all the methods
previously described for its solution have a rate of growth of the
computation time which is exponential in n (the number of cities
in the TSP). There are several approximation algorithms whose
rate of growth of the computation time is a low order polynomial in
n and that have been experimentally observed to perform well. In
this section we summarise some of these procedures, and restrict
our attention to symmetric TSPs with cost matrices that satisfy
the triangle inequality.

7.1 The Nearest Neighbour Rule (NNR)

In this procedure, we start with an arbitrary vertex and proceed
to form a path by joining the vertex just added to its nearest
neighbouring vertex which is not yet on the path, until all
vertices are visited; in this case, the two end vertices of the
hamiltonian path are joined to form the TSP solution. Rosenkrantz,
Stearns and Lewis [47] have shown that

$$\frac{V(NNR)}{V(TSP)} \leq \frac{1}{2}(\lceil \log n \rceil + 1), \tag{26}$$

where V(NNR) is the value obtained by NNR, V(TSP) is the value of
the optimal solution to the TSP, and $\lceil x \rceil$ is the smallest integer
greater than or equal to x.

Note that the cause of the worst-case bound on V(NNR) (given by
(26)), being logarithmically increasing with n, is neither the fact
that the last added arc in NNR is too long, nor that the starting
vertex is chosen arbitrarily. The NNR requires $O(n^2)$ operations to
apply.

7.2 The Nearest Insertion Rule (NIR)

In this procedure, we start with a circuit Φ passing through a
subset of the set of vertices, and add sequentially into Φ vertices
not already in Φ, until Φ becomes hamiltonian. The vertex (x, say)
to be added next at some stage can be chosen to be that vertex (not
in Φ) nearest to any vertex already in Φ; and having chosen x, it

44

is inserted in that position of Φ which causes the least
additional cost. The circuit Φ can be initialised to be a self-
loop on an arbitrarily chosen vertex. It is shown in [47] that

$$\frac{V(NTR)}{V(TSP)} < 2,$$

and that this bound is asymptotically tight.

Many possible variants of this algorithm come immediately to
mind, based on different ways of choosing the vertex x to be
inserted next, and choosing the position for its insertion. All
these variants have the same worst case bound mentioned above.
Depending on the variant, this algorithm requires $O(n^2)$ or
$O(n^2 \log n)$ operations to apply.

7.3 Lin's r-Optimal Heuristic [42]

Starting from an arbitrary initial tour, let r arcs be removed,
thus producing r disconnected paths. These paths can be reconnected
in one or more different ways to produce another tour, and this
operation is called an r-change. A tour is r-optimal if no r-change
produces a tour of lower cost.

It is shown in [47] that, for all $n \geq 8$, there exists a graph
for which

$$\frac{V(r-opt)}{V(TSP)} = 2(1 - n^{-1})$$

for all $r \leq \tfrac{1}{4}n$, where V(r-opt) is the value of an r-optimal tour.

Although the problem of finding an r-optimal tour can be performed in a number of operations polynomial in n, this number is
exponential in r and is bounded from below by n^r. Thus, only very
small values of r can be used in any heuristic [42], [13].

7.4 Christofides' Heuristic (CH) [14]

Let T* be the solution to the SST of graph G. Relative to T*, let
X° be the set of vertices having odd degree. Solve the matching
(1-matching) problem for graph $\langle X° \rangle$, and let M be the set of arcs
in this matching. The graph $G' = (X, A')$, composed from the set X

of vertices of G and having as arcs only those arcs in T* and M,
is eulerian; it can therefore be traversed by an eulerian circuit
E so that every arc of G' is traversed once and once only by E.
It is possible to construct a hamiltonian circuit Φ of G which
serves as the heuristic solution to the TSP by making use of
circuit E as follows:

Start constructing Φ by following arcs of E. If a vertex
already visited by Φ reappears in the vertex sequence of E, skip
that vertex unless all vertices are in Φ, in which case return to
the starting vertex.

It is shown in [14] that:

$$\frac{V(CH)}{V(TSP)} < \frac{3}{2},$$

and that CH requires $O(n^3)$ operations to be executed. Cornuejols
and Nemhauser [23] have shown that, for every $n \geq 3$, there exists
a graph for which:

$$\frac{V(CH)}{V(TSP)} = \frac{3n-1}{2n}.$$

The results given for the heuristics in Sections (7.1) to (7.4)
above are concerned with absolute performance guarantees. The
average performance of these heuristics is much better than the
worst case performance, and almost always within a few percent of
the optimum solution. This is particularly true for the r-optimal
heuristic of Lin and its variants [42], [13] (for r = 2 or 3),
which produces excellent results on average.

REFERENCES

1. A. Ainouche,   Hamiltonian circuits in graphs, Ph. D. thesis,
                      University of London, 1980.

2. A. Ainouche and N. Christofides, Strong sufficient conditions
                      for the existence of Hamiltonian circuits in
                      undirected graphs, J. Combinatorial Theory
                      (to appear).

3. E. Balas and N. Christofides, A new penalty method for the
                      travelling salesman problem, 9th Math. Prog.
                      Symposium, Budapest, 1976.

4. R. Bellman,   On a routing problem, Quart. J. Appl. Math. 16 (1958), 87-90.

5. M. Bellmore and J.C. Malone, Pathology of traveling-salesman subtour-elimination algorithms, Operations Res. 19 (1971), 278-307.

6. J.A. Bondy, and V. Chvátal, A method in graph theory, Discrete Math. 15 (1976), 111-135.

7. R.E. Burkard,   Travelling salesman and assignment problems — a survey, Report 77-11, Mathematisches Institut, University of Cologne, 1977.

8. P. Camerini, L. Fratta and F. Maffioli, Travelling salesman problem: Heuristically guided search and modified gradient techniques, Report of Instituto di Elettronica, Politecnico di Milano, 1975.

9. G. Carpaneto and P. Toth, An efficient algorithm for the asymetric TSP, ORSA National Meeting, Atlanta, November 1977.

10. N. Christofides, The shortest Hamiltonian chain of a graph, SIAM J. Appl. Math. 19 (1970), 689-696.

11. N. Christofides, Bounds for the travelling-salesman problem, Operations Res. 20 (1972), 1044-1056.

12. N. Christofides, Graph Theory — An Algorithmic Approach, Academic Press, London, 1975.

13. N. Christofides and S. Eilon, Algorithms for large-scale travelling salesman-problems, Operational Res. Quart. 23 (1972), 511-518.

14. N. Christofides, Worst case analysis of a new heuristic for the travelling salesman problem, Math. Prog. (to appear).

15. N. Christofides and C. Whitlock, An LP-based TSP algorithm, Imperial College Report 78-79, 1978.

16. N. Christofides, A. Mingozzi and P. Toth, Exact algorithms for vehicle routing, Math. Prog. (to appear).

17. N. Christofides, A. Mingozzi and P. Toth, The vehicle routing problem Combinatorial Optimisation (ed. N. Christofides et al.), John Wiley and Sons, New York, 1979.

18. N. Christofides, A. Mingozzi and P. Toth, State space relaxation procedures for the computation of bounds to routing problems, Networks (to appear).

19. N. Christofides and A. Ainouche, Hamiltonian circuits in graphs with large degrees, Imperial College Report, May 1980.

20. N. Christofides and A. Ainouche, Hamiltonian circuits in graphs: A survey, Imperial College Report, May 1980.

21. V. Chvátal,   Edmonds polytopes and weakly hamiltonian graphs, Math. Prog. 5 (1973), 29-40.

22. V. Chvátal and P. Erdös, A note on Hamiltonian circuits, Discrete Math. 2 (1972), 111-113.

23. G. Cornuejols and G.L. Nemhauser, Tight bounds for Christofides' TSP heuristic, Math. Prog. 10 (1978), 163.

24. H. Crowder and M. Padberg, The solution of large scale TSP's via linear programming, IBM Yorktown Heights Report, 1979.

25. G. D'Atri,   Improved lower bounds to the travelling salesman problem, IP77/3, Institut de Programmation, 1977.

26. G. Dantzig, D.R. Fulkerson and S.M. Johnson, Solution of a large-scale traveling-salesman problem, J. Operations Res. Soc. Amer. 2 (1954), 393-410.

27. J. Edmonds,   Maximum matching and a polyhedron with 0,1-vertices, Nat. Bur. Standards (B) 69B (1965), 125-130.

28. J. Edmonds,   Optimum branchings, J. Res. Nat. Bur. Standards (B) 71B (1967), 233-240.

29. D.R. Fulkerson, Packing routed directed cuts in a weighted directed graph, Math. Prog. 6 (1974), 1-13.

30. E.J. Gabovič,   The traveling salesman problem, Trudy Vyčisl. Centra Tartu. Gos. Univ. 19 (1970), 52-96.

31. M.R. Garey, R.L. Graham and P.S. Johnson, Some NP-complete geometric problems, Proc. Eighth ACM Symp. Theory of Computing, 1976.

32. P.C. Gilmore and R.E. Gomory, A solvable case of the traveling salesman problem, Proc. Nat. Acad. Sci. U.S.A. 51 (1964), 178-181.

33. M. Grötschel,   An optimal tour through 120 cities in Germany, Report 7770, University of Bonn, 1977.

34. M. Grötschel and M.W. Padberg, Linear characterisation of the symmetric travelling salesman polytope, Report 7417, University of Bonn, 1974.

35. K.H. Hansen and J. Krarup, Improvements of the Held-Karp algorithm for the symmetric travelling salesman problem, Math. Prog. 6 (1974), 87.

36. M. Held and R.M. Karp, The traveling-salesman problem and minimum spanning trees, Operations Res. 18 (1970), 1138-1162.

37. M. Held and R.M. Karp, The traveling-salesman problem and minimum spanning trees II, Math. Prog. 1 (1971), 6-25.

38. M. Held, P. Wolfe and H.P. Crowder, Validation of subgradient optimization, Math. Prog. 6 (1974), 62-88.

39. D. Houck, J-C. Picard, M. Queyranne and R.R. Vemuganti, The traveling salesman problem and shortest n-paths, Univ. of Maryland, 1977.

40. R.M. Karp, The probabilistic analysis of some combinatorial search algorithms, Proc. Symp. Algorithms and Complexity, Carnegie-Mellon Univ., Pittsburgh, (1976), 1-19.

41. E. Lawler, J.K. Lenstra and A. Rinnoy Kan, Scheduling: Complexity and Algorithms (to appear).

42. S. Lin, Computer solutions of the traveling salesman problem, Bell System Tech. J. 44 (1965), 2245-2269.

43. S. Lin and B.W. Kernighan, An effective heuristic algorithm for the traveling salesman problem, Operations Res. 21 (1973), 498.

44. P. Miliotis, Integer programming approaches to the travelling salesman problem, Math. Prog. 10 (1976), 367-378.

45. G.L. Nemhauser, L.A. Wolsey and M.L. Fischer, An analysis of approximations for maximising submodular set functions, CORE paper: 7618, Louvain, 1976.

46. O. Ore, Note on Hamilton circuits, Amer. Math. Monthly 67 (1969), 55.

47. D.J. Rosenkrantz, R.E. Stearns and P.M. Lewis, Approximate algorithms for the traveling salesperson problem, Proc. 15th IEEE Symp. Switching and Automata Theory (1974), 33-42.

48. M.I. Rubinšteǐn, On the symmetric traveling salesman problem, Automat. i Telemeh. (1971), 126-133.

49. C. Sandi, Subgradient optimisation, Combinatorial optimisation (ed. N. Christofides et al.), John Wiley and Sons, New York, 1979.

50. S. Sahni and T. Gonzalez, P-complete approximation problems, J. Assoc. Comput. Mach. 23 (1976), 555-565.

51. T.H.C. Smith and G.L. Thompson, A comparison of two different lagrangian relaxations of the TSP, MSR382, Carnegie-Mellon University, 1975.

52. M.M. Syslø, A new solvable case of the traveling salesman problem, Math. Prog. 4 (1973), 347-348.

N. Christofides
Department of Management Science
Imperial College of Science and Technology
Exhibition Road
London SW7 2BX

4 Graph theory and geography: some combinatorial and other applications
ANDREW CLIFF and PETER HAGGETT

1. <u>INTRODUCTION</u>

Since Leonhard Euler's study in 1736 of the seven bridges of
Königsberg, possible links between graph theory and geographical
situations have been apparent. Despite some continued interest in
the following century, as when Arthur Cayley published a note on
the four-colour problem in 1879, the main work on geographical
applications of graph theory has been of much more recent vintage,
dating mainly from about 1960. The publication of reports in
human geography by Garrison and Marble [7] and Kansky [12] on the
structure of transportation networks using elementary ideas of
graph theory was matched by a parallel interest in physical geography in the topological structure of stream networks (see [15],
[17], [18], [19]). Much of the work in the 1960s concentrated on
descriptive measures, whereas the more complex areas of the combinatorial structure of graphs and of algorithms for finding
optimal network flows have been explored only more recently.

In this paper, we begin by setting up a taxonomy of graphs in
tabular form. The structure of this table is used to lead the
discussion in the remainder of the paper. Throughout, the
emphasis is on giving representative examples of the work under-
taken in geography; for more comprehensive accounts with substan-
tial bibliographies, see the early text by Haggett and Chorley
[10] and the more recent reviews by Tinkler [23], [24] and by
Cliff, Haggett and Ord [3].

2. A TAXONOMY OF GRAPHS

A graph is a representation of a network that can be used as a
means of studying its structural properties. In its simplest
form, this consists of two sets of elements, the vertices and the
edges, and the only information we have about any edge or vertex
is its existence (1) or its absence (0). By contrast, a complex
graph will have information about the direction and values
(capacity, flow, frequency, etc.) of any edge and a separate set
of information about any vertex.

The following table presents the relationship between the level
of information available about edges and vertices in order to
build up a taxonomy of graphs. Data may be available at three
different levels:

(1)   dichotomous (presence or absence);

(2)   categorical (dichotomous, polychotomous, or ordered data on
the edges or vertices present in the graph);

(3)   continuous (interval or ratio data on the edges or vertices
present in the graph).

EDGE / VERTEX	PRESENCE OR ABSENCE	INFORMATION LEVEL ON EDGES	
		CATEGORICAL	CONTINUOUS
PRESENCE OR ABSENCE	CELL I	CELL II(a)	CELL III(a)
INFORMATION LEVEL ON VERTICES — CATEGORICAL	CELL II(b)	CELL II(c)	
INFORMATION LEVEL ON VERTICES — CONTINUOUS	CELL III(b)		CELL III(c)

Table 1

The table is organized in such a way that as we move from Cell I
to higher-order cells, the level of information about elements in
the graph increases; that is, we move from an abstract graph _sensu
strictu_ towards the sorts of real-world networks studied by
geographers, as illustrated by journey-to-work trips from origin
areas to destination areas within a city. The sequence is useful

52

in so far as it links those areas of quantitative spatial analysis which are conventionally labelled as graph theory, to those which are discussed in the geographical literature under such headings as integer programming or maximum-entropy modelling. In the following sections, we will adopt the sequence in the above table as a framework for discussion. For Cells I and II, we give some examples of the contributions made by geographers. Limitations on space prevent a similar treatment being given to Cell III and, in that case, appropriate references are given to the available literature.

3. ELEMENTARY GRAPHS

Much of the geographical work on elementary graphs (Cell I in the table) has focussed upon the physical interpretation of the eigenvalues and eigenvectors of transportation networks whose topological structure is presented in the form of an adjacency or shortest-path matrix [8]. A convenient algorithm for obtaining the eigenvalues and eigenvectors of a matrix, and one which is highly suggestive of the physical interpretation of these quantities, has been described by Kendall [13, pp.19-26] and extended by Cliff and Ord [4]. Given any matrix with real eigenvalues, this algorithm will, in general, extract these eigenvalues in order of descending absolute value. If $\underline{A}$ is the matrix in question, then corresponding to each eigenvalue λ_j is an eigenvector $\underset{\sim}{v}_j$ of unit length satisfying

$$\underline{A}\underset{\sim}{v}_j = \lambda_j \underset{\sim}{v}_j$$

The algorithm to extract these eigenvalues and eigenvectors has the following form:

Step 1. Choose $\underset{\sim}{v}^{(0)}$ to be an initial approximation to an eigenvector. (Here, it is adequate to choose $\underset{\sim}{v}^{(0)}$ to be $(1, 1, \ldots, 1)^T$. Scaling the largest component to unity is convenient during the iterative cycle, and quicker than scaling the sum of squares to unity, provided that we check that the largest element is not too close to zero before scaling.)

<u>Step 2</u>. For i = 1, 2, 3, ..., let $v^{(i)} = Av^{(i-1)}$, and continue
until the solution converges.

Once the first eigenvector v_1 has been found and normalized, we
form the new matrix $A^{(1)} = A - \lambda_1 v_1 v_1^T$, where λ_1 is the eigenvalue
corresponding to v_1. We can then repeat the above steps to obtain
the second eigenvector, and continue in this way until all of the
eigenvectors are known. (Further details are given in Cliff,
Haggett and Ord [3].)

The value of this algorithm in analyzing graphs was considered
by Tinkler [20]. The algorithm makes clear that the components of
the first eigenvector are directly proportional to the sum of the
entries in the corresponding row or column of the matrix under
consideration, and the eigenvalue is directly proportional to the
largest row or column sum. This implies that when a network is
being analyzed by means of its adjacency matrix, the more highly
connected the network is, the larger are its eigenvalues; and the
higher the valency of each vertex, the larger are the components
of the corresponding eigenvector. However, if the shortest-path
matrix is analyzed, the good overall connectedness of the graph is
reflected in a low first eigenvalue, and well-connected individual
vertices have low eigenvector components. Badly-connected networks
and vertices exhibit the reverse characteristics – namely, low
eigenvalues and eigenvector components for the adjacency matrix,
and high eigenvalues and eigenvector components for the shortest-
path matrix.

The fact that the largest eigenvalue is a measure of the overall
connectedness of a transportation network can be established by
defining bounds on it (see Cliff and Ord [4]). If λ_1 is the
largest eigenvalue of the adjacency matrix $A = (a_{ij})$, then

$$\frac{1}{n} \sum_{i,j} a_{ij} \leq \lambda_1 \leq \max_i \sum_j a_{ij} .$$

(The lower bound is simply the average vertex-degree, and the
upper bound is the maximum vertex degree.) Similarly, if $S = (s_{ij})$
is the shortest-path matrix, then

$$\frac{1}{n} \sum_{i,j} s_{ij} \leq \lambda_1 \leq \max_i \sum_j s_{ij} \; .$$

(Here the lower bound is the average number of edges travelled per journey, and the upper bound is the longest journey in the network.)

The physical interpretation of the second and subsequent eigenvalues and eigenvectors has been the subject of some controversy (see, for example, Hay [11] and Tinkler [22]). However, there is some evidence to suggest that the out-valencies of the vertices may be used to separate out well-connected subgraphs within the main graph.

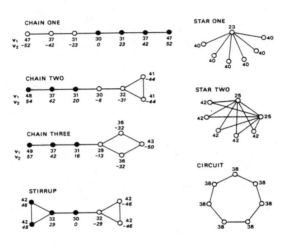

Figure 1

In order to illustrate these points, we have computed the eigenvalues and eigenvectors of the adjacency and shortest-path matrices of several artificial networks (Figure 1), and of the airline networks connecting the seven largest cities in four countries (Figure 2). In these diagrams we have given next to each vertex the components of the eigenvector corresponding to the largest (Roman numbers) and, in most cases, the second-largest (italicised numbers) eigenvalue of $\underline{S}$ in absolute value. (In each case the eigenvector is chosen to have length 100.) In general, the larger the degree of the vertex, the lower is the corresponding

component of the first eigenvector. In the networks of Figure 1,
the components of the second eigenvector seem to delineate distinct
regions, with the 'between-regions' links being relatively weak
compared to the 'within-region' links.

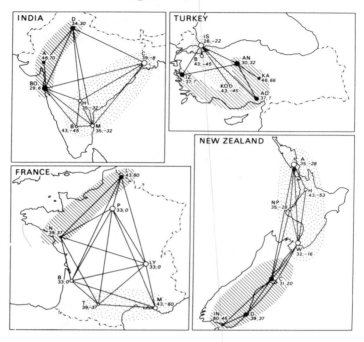

Figure 2

In Figure 2, the regions which arise are indicated by shading
the corresponding groups of vertices; the sizes of the vertices
indicate the ranking of the cities according to population size.
Multiple eigenvalues tend to occur whenever two vertices are
'interchangeable' — that is, when they are joined to the same
vertices. For such networks the second eigenvector is meaningless.
Although empirical networks are less prone to give rise to multiple
eigenvalues, even here eigenvalues of similar magnitude may arise,
thereby making regional interpretations rather difficult.

4. DIRECTED, CHROMATIC AND ORDERED GRAPHS

The second tier of cells in the above table includes graphs where
either the edges [II(a)], or the vertices [II(b)], or both [II(c)],

are labelled with information measured in a categorical manner.
The categories may be dichotomous (such as the direction of flow
along an edge), or polychotomous (for example, vertices might be
coloured according to some categorization), or ordered (where the
several categories are in a fixed relationship to each other).
Graphs labelled in these three ways are termed directed graphs,
chromatic graphs, and ordered graphs, respectively. Each of the
three types may occur in any of the three cells.

Cell II(a) : streams as hierarchical and random graphs

Apart from any section forming a closed loop, stream channels show
all the topological properties of a rooted tree. Such branching
graphs have attracted the attention of earth scientists for the
last half-century, as part of their search for answers to questions about stream evolution and the relationship between the
morphology of a stream network and its hydrology (for example, the
flood characteristics). One significant research area has been
the development of random topology models of stream systems based
on their combinatorial properties. Under this approach, each
channel is assigned a number called its magnitude, defined to be
the total number of sources upstream. Thus the magnitude of each
edge is the sum of the magnitudes of the edges leading into it.
The magnitudes of stream channels have been used by Strahler [19]
to define the order of a stream system. Channels with magnitude 1
are called 'first-order channels'. Where these merge, a second-
order channel is formed; where two second-order channels merge, a
third-order channel is formed, and so on. The highest order
stream segment fixes the order of the system. The amount of
branching in the system is summarized in the bifurcation ratio.
If N_i is the number of streams of order i, the i-th bifurcation
ratio R_i is defined as $R_i = N_i/N_{i+1}$, and the average bifurcation
ratio $\bar{R}$ is the average over all $\{R_i\}$ for the stream system under
consideration.

Stream networks with the same magnitude may have different
topological structures, as illustrated in Figure 3. This shows

the forty-two topologically distinct channel networks (TDCNs) with
six sources and eleven links [3]. If there are k sources, then
the number $N(k)$ of TDCNs is given by the Catalan number

$$N(k) \ = \ \frac{1}{2k-1} \binom{2k-1}{k} \ .$$

For networks with one to nine sources, these numbers are
1, 1, 2, 5, 14, 42, 132, 429 and 1430, respectively.

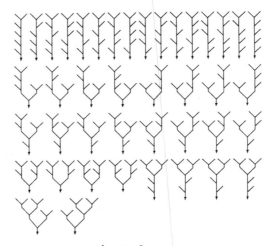

Figure 3

Given the large number of possible TDCNs, geographers have been
interested in two related questions:
(i) which is the 'most probable' pattern?
(ii) do naturally-occurring stream networks approximate to this
most probable pattern?

Consideration of Figure 3 shows that it is possible to divide
the forty-two TDCNs there into two groups in terms of their
Strahler-order — namely, sixteen second-order networks (in the
first row), and twenty-six third-order networks (in rows 2-5).
The following table classifies the TDCNs with $k \leq 9$ according to
their Strahler-order. Thus for $k = 9$, a third-order network is
the most likely type, with a probability of $\frac{1288}{1430}$ ($\cong 0.90$), whereas
the second-order and fourth-order systems have probabilities of

about 0.09 and 0.01, respectively.

number of sources k	number of TDCNs with Strahler-order				total number of TDCNs
	1	2	3	4	
1	1				1
2		1			1
3		2			2
4		4	1		5
5		8	6		14
6		16	26		42
7		32	100		132
8		64	364	1	429
9		128	1288	14	1430

With a large number of sources, the number of TDCNs becomes very large indeed. For example, a stream system with fifty sources has 5.10×10^{26} possible patterns. At one extreme, the fifty first-order streams might be flowing into a single elongated second-order channel to give the highly improbable average bifurcation ratio, $\bar{R} = 50$. At the other extreme, we can envisage a highly-branched structure with $\bar{R} \cong 2.22$, arising from a sixth-order system with a sequence of 50, 25, 12, 6, 3, 1 stream segments of ascending Strahler-order. Werrity [25] has shown that the most probable stream number set with fifty sources has the sequence 50, 12, 3, 1, giving a fourth-order system with $\bar{R} = 3.83$; the probability of such a stream system arising is approximately 0.097. Fourth-order networks of all kinds make up the overwhelmingly likely set (about 60%) of all possible stream systems with fifty sources.

Empirical studies from a number of different climatic and geological environments show a clustering of bifurcation ratios in the range 3.5 to 4.0. One of the most interesting results of such work on random networks is that the most probable network of a given order is that in which the bifurcation ratio R_1 is close to 4.0. Werrity's detailed fieldwork in south-west England has confirmed the general agreement between the observed topological characteristics of stream systems and those predicted by the random TDCN model. Divergences from the random model, where they occur, can be attributed to specific geological controls or erosional history. Thus we can conclude that most stream systems have a branching structure which satisfies the most probable TDCN with $R_1 \cong 4.0$.

Cell II(b) : spatial diffusion in chromatic graphs

Insights into the structure of diffusion processes can often be gained by reducing the map to a graph in which the areas become the vertices and the links between them become the edges. The vertices are then appropriately two-coloured depending upon whether or not an event has occurred at a given vertex. Such an approach was taken by Haggett [9] in an analysis of measles epidemics in Cornwall. Haggett constructed seven different graphs of the 27 local authority (General Register Office) areas in Cornwall; each graph was designed to correspond as closely as possible with a hypothetical diffusion process, namely:

G_1: local contagion with the assumption of spread only between contiguous GRO areas — planar graph with 34 edges.

G_2: wave contagion with the assumption of spread by shortest paths from an endemic reservoir area (Plymouth)— planar graph with 28 edges.

G_3: regional contagion with the assumption that spread is locally contagious and not on a county-wide basis, but rather within two separate regional subsystems (east and west Cornwall)— planar subgraphs with 32 edges.

G_4: urban-rural contagion with the assumption of spread within sets of urban and rural communities treated as separate subgraphs— non-planar subgraphs with 181 edges.

G_5: population size with the assumption of spread down the hierarchy of population size from largest to smallest centre— non-planar graph with 26 edges.

G_6: population density with the assumption of spread through the density hierarchy — non-planar graph with 26 edges.

G_7: journey-to-work contagion with the assumptions that these flows

60

provide a surrogate for spatial interaction between areas and that spread follows high interaction links — non-planar graph with 58 edges.

To discriminate between the seven graphs, all the 222 weekly maps of measles outbreaks studied by Haggett were translated into outbreak/no outbreak terms. The vertices of each of the 222 × 7 graphs were colour-coded black (B = outbreak) or white (W = no outbreak) and the Moran BW statistic [14] under non-free sampling was evaluated to measure the degree of contagion present in the graph. The greater the degree of correspondence between any graph and the transmission path followed by the diffusion wave, the larger (negative) should become the standard score for BW. In practice, there may be practical problems of common links as discussed in Haggett [9, p.145].

As a result of testing the spread patterns on the seven graphs G1-G7, the following preliminary observations can be made.
(1) The spatial diffusion process was readily separated into inter-epidemic and epidemic periods, with different spread processes dominant in each.
(2) During the long periods between main epidemics there was a lower level of contagion on the graphs based upon spatial structure (G_1-G_4 and G_7), so that any diffusion currents recorded were extremely weak and poorly structured. Infections persisted in the larger population clusters (G_5, G_6) and moved slowly through the low-density rural areas in a sporadic manner.
(3) During the shorter epidemic periods, the general level of contagion was about $1\frac{1}{2}$ times higher, so that spatial processes were more distinctive and easier to monitor. The advance phase of the epidemic was marked by a rapid increase both in intensity and spread. Population size became a less important determinant and wave and town-country effects increased sharply (G_2, G_4). At the peak of the epidemic wave, local contagion and regional effects became important, setting up strong areal contrasts between compact clusters of infected/non-infected areas (G_1, G_3). During the subsequent retreat phase, the falloff in intensity was not associated with a corresponding reduction in geographical area, so that the

epidemic appeared to decay spatially _in situ_. Contagion in G_1, G_2 and G_3 fell steadily during the retreat phase: one exception was the spatial interaction model G_7, which showed somewhat higher values for a greater length of time after the peak; this implies that longer-range contacts between population centres may come later in the history of the epidemic wave.

Although this suggested history of the diffusion process of an epidemic must remain speculative for the reasons given in Cliff and Haggett [2], this analysis of a spread process using chromatic graphs identifies for a childhood disease the hierarchical and contagious elements generally associated with innovation diffusion. It further indicates that there is a strong contrast between the early build-up (population size dominates) and fade-out phases (spatial structure dominates) of an epidemic.

Cell II(c) : market cycles as ordered graphs

One major area of geographical research concerned with the ordered structure of human settlements is 'central-place theory'. Settlements may be conceived as an irregular planar lattice in which functions are provided on a permanent or periodic basis. For example, rural centres may hold markets on different days of the week. It will be clear (see Figure 4) that such periodic functions may form a 'market-cycle' in time and a 'market ring' in space. In the simple example shown, the six smaller settlements form a regular once-a-week cycle, with a seventh vacant day for rest or restocking: arrows indicate the ring in space formed by the presumed movement of an individual trader. Empirical regional studies show a wide variety of cycles which are unrelated, or only loosely related, to the conventional seven-day week. Two-day and four-day systems have been described in East and West Africa, and ten-day cycles have been noted in China.

Tinkler [21] has approached the periodic market as a map-colouring problem in which no market towns in adjacent counties can have the same colour (market day). Obviously, direct competition would be self-defeating for the markets in such a situation, and it follows from the four-colour theorem that four days will ensure that there are no conflicts between adjacent markets.

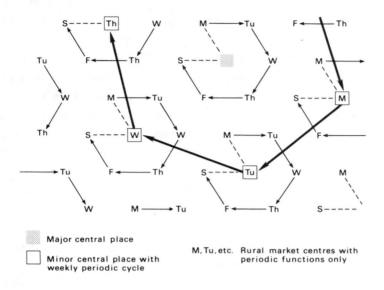

Figure 4

Figure 5 shows some possible four-day periodic systems. Note that we can conceive of both a closed system (A) and an open system (B), but that greater advantages (in terms of exchange of goods) accrue under the open system; (C) shows how higher and lower order periodic systems may be linked together.

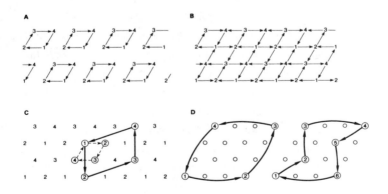

Figure 5

One further advantage of the four-day cycle is that it is
relatively easy to modify it to a six-day cycle by including new
markets within the centres of the triangles in the lattice structure (D). With this modification, it fits readily into the six-
working day calendar common in most countries. As a counter-
balance to these theoretical ring studies, Skinner [16] summarizes
the main features of a ten-day cycle for rural areas in the
Szechwan province of China. The resilience of these market
structures is attested by the breakdown of the more centralized
state trading centres superimposed on this area in the early 1950s,
and the total re-establishment of the traditional market cycle a
decade later.

5. CONCLUSION

The outer tier of cells in the first table spans those graphs
where the information loaded on edges [III(a)], vertices [III(b)],
or both [III(c)], is measured on interval or ratio scales. Each
of the three cells is well represented in quantitative geography
and they overlap with research problems on which whole books have
been written. For example, in Cell III(a) Forer [6] has used cost
and time information on edges to produce space-time maps of the
changing cost and time distances by air between places in New
Zealand and the Pacific. Likewise, in Cell III(b) Cliff and Ord
[5] have used connection matrices as a basis for spatial auto-
correlation and correlogram analysis. Finally, Cell III(c)
problems, where both edges and vertices are fully measured, are
fundamental inputs to gravity models, maximum entropy models, and
linear programming (transportation problem) models; standard
references are those of Wilson [26] and Batty [1].

Because of the tendency of graphs (as a spatial form) to dissolve
into aspatial matrices, they are sometimes overlooked as a distinctive phenomenon for quantitative geographical research. In this
paper we have argued for a taxonomy of graphs which runs from their
familiar and overt elementary form in the analysis of phenomena such
as stream channels and transport networks to their embedded and
covert role in gravity, entropy maximizing and linear programming

64

models. Geographical examples of both methods of use have been provided.

As we indicated in our introduction, the links between graph theory and geography are long-standing. The illustrations given in this paper show some areas where limited progress has been made, and indicate others where further work is continuing. In general, combinatorialists have been unaware of the kinds of problems on which geographers are working, while geographers are conversant with only the more elementary of graph theory concepts. The potential for further collaboration between the two is clearly considerable.

REFERENCES

1. M. Batty,   Urban Modelling: Algorithms, Calibrations, Predictions, Cambridge University Press, Cambridge, 1975.

2. A.D. Cliff and P. Haggett, Geographical aspects of epidemic diffusion in closed communities, Statistical Applications in the Spatial Sciences (ed. N. Wrigley), Pion, London (1979), 5-44.

3. A.D. Cliff, P. Haggett and J.K. Ord, Graph theory and geography, Applications of Graph Theory (ed. R.J. Wilson and L.W. Beineke), Academic Press, London (1979), 293-326.

4. A.D. Cliff and J.K. Ord, Latent roots and vectors of an arbitrary real matrix, Environment and Planning (A) 9 (1977), 703-714.

5. A.D. Cliff and J.K. Ord, Spatial Processes, Pion, London, 1981.

6. P. Forer,   Changes in the Spatial Structure of the New Zealand Airline Network, Ph.D. thesis, University of Bristol, 1974.

7. W.L. Garrison and D.F. Marble, The structure of transportation networks, U.S. Army, Transportation Research Command Contract 44-177-TC-685, Task 9R98-09-003-01, Technical Report 62-11, 1962.

8. P.R. Gould,   On the geographical interpretation of eigenvalues, Inst. Brit. Geog. Publ. 42 (1967), 63-92.

9. P. Haggett,   Hybridizing alternative models of an epidemic diffusion process, Econ. Geog. 52 (1976), 136-146.

10. P. Haggett and R.J. Chorley, Network Analysis in Geography, Edward Arnold, London, 1969.

11. A. Hay,   On the choice of methods in the factor analysis of connectivity matrices: a comment, Inst. Brit. Geog. Publ. 66 (1975), 163-167.

12. K.J. Kansky,   Structure of transport networks: relationships between network geometry and regional characteristics, Department of Geography Research Papers 84, Univ. of Chicago, 1963.

13. M.G. Kendall,   Multivariate Analysis, Griffin, London, 1957.

14. P.A.P. Moran,   The interpretation of statistical maps, Journal of the Royal Statistical Society 10 (1948), 243-251.

15. R.L. Shreve,   Statistical law of stream numbers, J. Geology 74 (1966), 17-37.

16. S.W. Skinner,   Marketing and social structure in rural China, J. Asian Studies 24 (1964-65), 3-399.

17. J.S. Smart,   A comment on Horton's law of stream numbers, Water Resources 4 (1967), 1001-1014.

18. J.S. Smart,   Topological properties of channel networks, Bull. Geolog. Soc. America 80 (1969), 1757-1774.

19. A.N. Strahler,   Quantitative geomorphology of drainage basins and channel networks, Handbook of Applied Hydrology (ed. V.T. Chow), McGraw-Hill, New York (1964), 4.40-4.74.

20. K.J. Tinkler,   The physical interpretation of eigenvalues of dichotomous matrices, Inst. Brit. Geog. Publ. 55 (1972), 17-46.

21. K.J. Tinkler,   The topology of rural periodic market systems, Geografiska Annaler (B) 55 (1973), 121-133.

22. K.J. Tinkler,   On the choice of methods in the factor analysis of connectivity matrices: a reply, Inst. Brit. Geog. Publ. 66 (1975), 168-171.

23. K.J. Tinkler,   An introduction to graph theoretical methods in geography, Concepts and Techniques in Modern Geography 14 (1977).

24. K.J. Tinkler,   Graph theory, Progress in Human Geography 3 (1979), 85-116.

25. A. Werrity,   The topology of stream networks, Spatial Analysis in Geomorphology (ed. R.J. Chorley), Methuen, London (1972), 167-196.

26. A.G. Wilson,   Urban and Regional Models in Geography and Planning, John Wiley and Sons, London, 1974.

Andrew D. Cliff   Peter Haggett
Department of Geography   Department of Geography
Downing Place   University Road
Cambridge CB2 3EN   Bristol BS8 1SS

5 Coding, decoding and combinatorics
J.H. VAN LINT

1. INTRODUCTION

This is an expository lecture; most of what we describe can be
found in [1] and [2]. As a starting point we take the BBC/OU
television programme [3][†] about the Mariner 9 mission. We shall
consider in more detail the contribution of combinatorial theory
to the success of the coding and decoding schemes. We then consider generalisations of the codes used by Mariner 9 based on
affine finite geometries.

We first remind the reader of the way in which pictures taken
by satellites are transmitted to earth. A fine grid is placed on
the picture, and for each square of the grid the degree of blackness is measured on a scale from 0 to 63. These numbers are
expressed in the binary system, i.e. each square produces a
sequence of zeros and ones. The zeros and ones are transmitted as
two different signals to the receiver station on earth (the Jet
Propulsion Laboratory of the California Institute of Technology in
Pasadena). On arrival the signal is very weak and it must be amp-
lified. Noise added to the signal and thermal noise from the amp-
lifier have the effect that it happens occasionally that a signal
which was transmitted as a zero is interpreted by the receiver as
a one, and vice versa. We shall denote the probability that this
happens by p (and we write q := 1 - p). If the sextuples of zeros
and ones which we mentioned above were transmitted as such, then

---

[†] This television programme, featuring Professor van Lint, was
shown at the conference before his talk. It describes the
American Mariner 9 mission to Mars, and discusses the problem
of finding a suitable code for relaying information back to
Earth.

the errors made by the receiver would have great effect on the
pictures. Assume for example that p = 0.05. Then the probability
that the degree of blackness of one square is interpreted correctly
by the receiver is $q^6 = 0.74$, i.e. 26% of the picture will be wrong!
In order to prevent this, redundancy is built into the signal, i.e.
the transmitted sequence consists of more than the necessary information. In the case of Mariner 9 it was acceptable to have the
transmission take about five times as long as would be necessary
without coding. Now suppose we repeat each bit of a sextuple
$\underline{a} := (a_0, a_1, \ldots, a_5)$ five times. A bit 0 would be transmitted as
00000 and the received fivetuple would be decoded into the symbol
occurring most often. In this case the probability that $\underline{a}$ is
interpreted correctly is $(q^5 + 5q^4 p + \binom{5}{2} q^3 p^2)^6 = 0.96$. The
probability of error has been reduced to 4%. But, with practically
the same rate of transmission, we can do much better. The solution
which was used is as follows.

Let H be a Hadamard matrix of size 32. We consider the rows of
H and $-$H. In each row we replace the symbols $+1$ and -1 by 0 and 1
respectively. In this way we obtain 64 <u>words</u> of length 32 which
are used as signals for the 64 degrees of blackness. This is the
so-called <u>code</u>. Since any two distinct rows of H or $-$H have inner
product 0 or -32, they differ either in 16 positions or everywhere.
The same holds for the codewords. It follows that if a word is
received with anywhere up to seven errors it is interpreted correctly. Therefore the probability of correct reception is

$$\sum_{k=0}^{7} \binom{32}{k} p^k q^{32-k} = 0.9999,$$

i.e. we now have a probability of error of $\frac{1}{100}$ %, a tremendous
improvement!

A simple scheme for decoding is as follows. A received signal,
i.e. a sequence of 32 zeros and ones, is changed into its ± 1 form
(replace 0,1 by 1,-1). If the result is $\underline{x}$ and if there are no
errors, then $\underline{x} H^T$ will be a vector with 31 components equal to 0
and one component equal to ± 32. In the presence of errors these

68

numbers are changed, but if the number of errors is at most 7 then the values 0 can increase to at most 14 and the value 32 can decrease to no less than 18. So the maximal entry in $\underline{x}\,H^T$ will tell us which row of H was transmitted (with the obvious modification if it was a row of $-H$).

2. CODING AND DECODING GAINS BY THE USE OF COMBINATORICS

It is desirable to have a simple algorithm which generates the signal for a given degree of blackness (satellites are small!). This is done by using special Hadamard matrices as follows. Let $H_2 := \begin{pmatrix} 1 & 1 \\ 1 & -1 \end{pmatrix}$. Define $H_4 := H_2 \otimes H_2$ and, in general,

$$H_{2^{n+1}} := H_2 \otimes H_{2^n} .$$

Consider a Hadamard matrix H_n, where $n = 2^m$, obtained in this way. Let C_n be the corresponding $(0,1)$-matrix. We claim that the 2^{m+1} rows of C_n and $J-C_n$ form a linear subspace of dimension $m + 1$ in $(\mathbb{F}_2)^n$. This is easily proved by induction. First, observe that the rows of $J-C_n$ are obtained from those of C_n by adding $\underline{1} := (1,1,\ldots,1)$. For $n = 2$ the assertion is obvious. Suppose the statement is true for n, and let $\underline{x}_1, \underline{x}_2, \ldots, \underline{x}_n$ be the basis vectors corresponding to C_n. Since $H_{2n} = H_2 \otimes H_n$ we have

$$C_{2n} = \begin{pmatrix} C_n & C_n \\ C_n & J-C_n \end{pmatrix}$$

which is clearly generated by the basis $(\underline{x}_1,\underline{x}_1)$, $(\underline{x}_2,\underline{x}_2)$, $\ldots$, $(\underline{x}_m,\underline{x}_m)$, $(\underline{0},\underline{1})$.

We shall denote the linear code of length $n = 2^m$ and dimension $m + 1$ obtained in this way by $\underline{R}(1,m)$. A basis for this code is

$$\underline{v}_0 := (1,1,1,\ldots,1),$$
$$\underline{v}_1 := (0,1,0,1,\ldots,0,1),$$
$$\underline{v}_2 := (0,0,1,1,\ldots,0,0,1,1),$$
$$\cdot \quad \cdot \quad \cdot \quad \cdot \quad \cdot$$
$$\underline{v}_m := (0,0,\ldots,0,1,1,\ldots,1).$$

In the case of Mariner 9 we had $m = 5$, $n = 32$. A sextuple $\underline{a} = (a_0, a_1, \ldots, a_5)$ is encoded as $\sum_{i=0}^{5} a_i \underline{v}_i$. This is easier to realize in hardware than it is to store the 64 rows of H and −H.

Now we turn to decoding. Again, the fact that a Kronecker product was used allows us to speed up decoding considerably. To see this we first consider the method described in Section 1. For a received word $\underline{x}$ (in its ±1 notation) we calculate $\underline{x} \, H_n^T$. This involves n^2 multiplications and the corresponding addition operations.

Define the matrix $M_{2^m}^{(i)}$ by

$$M_{2^m}^{(i)} := I_{2^{m-i}} \otimes H_2 \otimes I_{2^{i-1}} \qquad (1 \leq i \leq m).$$

For example,

$$M_8^{(2)} = I_2 \otimes H_2 \otimes I_2$$

$$= \begin{bmatrix}
+ & 0 & + & 0 & & & & \\
0 & + & 0 & + & & \mathbf{0} & & \\
+ & 0 & - & 0 & & & & \\
0 & + & 0 & - & & & & \\
& & & & + & 0 & + & 0 \\
& \mathbf{0} & & & 0 & + & 0 & + \\
& & & & + & 0 & - & 0 \\
& & & & 0 & + & 0 & -
\end{bmatrix},$$

(here + indicates 1, − indicates −1).
An easy induction proof shows that

$$H_n = M_n^{(1)} \, M_n^{(2)} \, \ldots \, M_n^{(m)} . \qquad (*)$$

To calculate the inner product of a vector with the rows of $M_n^{(i)}$ involves only $2n$ multiplications. So if we calculate $\underline{x} \, H_n^T$ by iteration, using (*), we need only $2nm$ multiplications. In the case of Mariner 9 this is about one third of the amount needed without this trick. A more detailed analysis shows that the work can be reduced by a factor $1/5$.

70

3. FINITE GEOMETRIES AND REED-MULLER CODES

Consider the affine space $AG(m,2)$. As before, let $n := 2^m$. We number the points of the space from 0 to $n - 1$ by giving $(x_1, x_2, \ldots, x_m)$ the number $\sum_{i=1}^{m} x_i \cdot 2^{i-1}$. The characteristic function of a subset S of $AG(m,2)$ is a $(0,1)$-vector $\underline{s}$ in F_2^n. We identify S and $\underline{s}$. For the basis vectors of $\underline{R}(1,m)$ given in Section 2 we see that $\underline{v}_0$ corresponds to $AG(m,2)$ itself, and for $1 \leq i \leq m$ the vector $\underline{v}_i$ is the characteristic function of the hyperplane $V_i := \{\underline{x} \in AG(m,2) \mid x_i = 1\}$. It follows that the code $R(1,m)$, which is called the first order Reed-Muller code of length $n = 2^m$, consists of the characteristic functions of $\emptyset$, $AG(m,2)$, and the $2n - 2$ hyperplanes of $AG(m,2)$. In fact,
$$a_0 \underline{v}_0 + a_1 \underline{v}_1 + \ldots + a_m \underline{v}_m$$ is the characteristic function of the hyperplane

$$\{\underline{x} \in AG(m,2) \mid a_0 + a_1 x_1 + \ldots + a_m x_m = 1\}$$

We now generalize. Let $\underline{v}_i \underline{v}_j$ $(1 \leq i < j \leq m)$ be the coordinate-wise product. Then this corresponds to the $(m-2)$-flat

$$\{\underline{x} \in AG(m,2) \mid x_i = 1 \text{ and } x_j = 1\}.$$

The second order Reed-Muller code $\underline{R}(2,m)$ of length $n = 2^m$ is the linear code of dimension $1 + m + \binom{m}{2}$ with $\underline{v}_i$ $(0 \leq i \leq m)$ and $\underline{v}_i \underline{v}_j$ $(1 \leq i < j \leq m)$ as basis vectors.

Our knowledge of finite geometries greatly facilitates the treatment of these codes. First, we observe that every codeword is apparently a linear combination of characteristic functions of hyperplanes and $(m-2)$-flats. Conversely, we claim that the characteristic function of an $(m-2)$-flat is a codeword in $\underline{R}(2,m)$. For, if V is an $(m-2)$-flat, then V is the intersection of two hyperplanes, say

$$\{\underline{x} \in AG(m,2) \mid a_0 + a_1 x_1 + \ldots + a_m x_m = 1\}$$

and $\quad \{\underline{x} \in AG(m,2) \mid b_0 + b_1 x_1 + \ldots + b_m x_m = 1\}.$

Hence

$$V = \{\underline{x} \in AG(m,2) \mid (a_0 + a_1x_1 + \ldots + a_m x_m)(b_0 + b_1x_1 + \ldots + b_m x_m) = 1\},$$

which is clearly a linear combination of the basis vectors.
Another way of looking at things is to observe that the codewords
correspond to <u>quadrics</u> in $AG(m,2)$. For example, $\underline{v_1}\underline{v_2} + \underline{v_3}\underline{v_4}$ in
$\underline{R}(2,4)$ corresponds to the quadric $\{\underline{x} \in AG(m,2) \mid x_1x_2 + x_3x_4 = 1\}$.

How can we use these codes in practice? It is not difficult to
show by induction that two distinct codewords in $\underline{R}(2,m)$ differ in
at least 2^{m-2} positions (i.e., the code has <u>minimum distance</u> 2^{m-2}).
Hence the code is capable of correcting up to $2^{m-3} - 1$ errors.

Suppose we receive a word $\underline{y} = (y_0, y_1, \ldots, y_{n-1})$ containing
$e \le 2^{m-3} - 1$ errors. Suppose there is an error in the position
corresponding to the point $\underline{x} \in AG(m,2)$. There are $2^m - 1$ lines in
$AG(m,2)$ through $\underline{x}$, and $e - 1$ of these 'contain' two errors; the
others (i.e. the majority) contain only one error. If the
position $\underline{x}$ does not correspond to an error then the majority of the
lines through $\underline{x}$ contain 0 errors (an even number). It follows
that if we know for each line of $AG(m,2)$ whether it contains an
even or an odd number of errors, then we can find and correct all
errors.

The procedure using this idea (<u>majority logic</u> decoding) starts
with the 3-flats of the geometry. We use the fact that we know
that every 3-flat intersects a hyperplane and an $(m-2)$-flat in an
<u>even</u> number of points. Therefore every codeword $\underline{c}$ must have an
even number of points in each 3-flat. Consider any plane
(= 2-flat) T. There are $2^{m-2} - 1$ distinct 3-flats F_i (i = 1, 2,
..., $2^{m-2} - 1$) containing T. Every point not in T is in exactly
one of the F_i. We now use the same principle as above. If $\underline{y}$
contains an odd number of errors in the positions of T, then the
majority of the numbers $|\underline{y} \cap F_i|$ will be odd (and similarly for
even). So a majority vote decides on the behaviour of T. We do
this for every plane, and then move down to lines and finally to
points. Success for geometry!

4. KERDOCK CODES

In this section m is even. In Section 3 we saw that in going from
$\underline{R}(1,m)$ to $\underline{R}(2,m)$ the dimension increased from $1 + m$ to $1 + m + \binom{m}{2}$
and the minimum distance went down from 2^{m-1} to 2^{m-2}. The Kerdock
codes which we now treat by example are in between. The code $\underline{K}(m)$
has 2^{2m} words (it is not linear) and minimum distance
$2^{m-1} - 2^{(m-2)/2}$ (i.e. only slightly less than $\underline{R}(1,m)$, although
$\underline{K}(m)$ has nearly twice the "dimension"). There are still no simple
descriptions of these codes, and several questions are not understood. This is a good area for further research.

The code $\underline{K}(4)$ will consist of $\underline{R}(1,4)$ and seven cosets of $\underline{R}(1,4)$
in $\underline{R}(2,4)$. The cosets are described by codewords in $\underline{R}(2,4)$ — that
is, by quadratic forms or, equivalently, by quadrics in $AG(4,2)$.
In this geometry every non–degenerate quadric is equivalent to

$$Q_0 := \{\underline{x} \in AG(m,2) \mid x_1 x_2 + x_3 x_4 = 1\}.$$

Every hyperplane intersects Q_0 in six or in ten points. This
means that the word $\underline{v}_1 \underline{v}_2 + \underline{v}_3 \underline{v}_4$ in $\underline{R}(2,4)$ has distance 6 or 10
from every codeword in $\underline{R}(1,4)$. So, we see that in order to con-
struct the code $\underline{K}(4)$ with minimum distance 6, we need seven quadratic forms $q_1, q_2, \ldots, q_7$ such that for every i and j ($i \neq j$) the
form $q_i + q_j$ corresponds to a non–degenerate quadric.

In order to find these we use the following compact notation.
Let $\underline{F}_4 = \{0, 1, \omega, \omega^2 = \omega + 1 = \bar{\omega}\}$. Since $\underline{F}_4$ is a 2–dimensional
vector space over F_2 we can interpret $(a,b,c) \in \underline{F}_4^3$ as a
$(0,1)$-vector in $\underline{F}_2^6$. Number the monomials $x_i x_j$ in the order
$x_1 x_2, x_3 x_4; x_1 x_3, x_2 x_4; x_1 x_4, x_2 x_3$. We denote a combination of
these by a vector $(a,b,c) \in \underline{F}_4^3$. For example, $(1,\omega,\bar{\omega})$ corresponds
to $x_1 x_2 + x_2 x_4 + x_1 x_4 + x_2 x_3$. Now it is easy to see that there are
28 combinations corresponding to non–degenerate quadrics and these
are given by (a,b,c), where one or three of the coordinates are
equal to $\bar{\omega}$. This easy representation allows one to pick the
required set of seven at a glance. We take $(\bar{\omega},\bar{\omega},\bar{\omega})$, $(\bar{\omega},0,1)$,
$(\bar{\omega},0,\omega)$ and their cyclic shifts. It is obvious that the sum of any
two has one coordinate $\bar{\omega}$.

At present the desciption of $\underline{K}(m)$, m even, m > 4, is considerably more difficult.

REFERENCES

1. J.H. van Lint,   Introduction to Coding Theory, Graduate Texts in Mathematics, Springer-Verlag, New York, 1982.

2. F.J. MacWilliams and N.J.A. Sloane, The Theory of Error-Correcting Codes, North-Holland, Amsterdam, 1977.

3.   The Mariner 9 Code, Television programme 14 of TM361: Graphs, Networks and Design, The Open University, Milton Keynes, 1981.

J.H. van Lint
Department of Mathematics
Teknische Hochschule
Eindhoven
The Netherlands

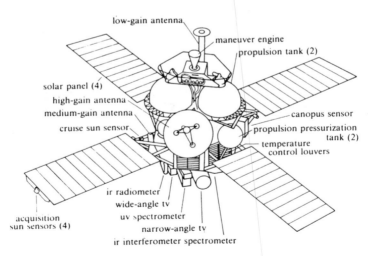

Mariner 9 spacecraft from above

6 Problems in computational complexity
D.J.A. WELSH

1. INTRODUCTION

The ideas and concepts of computational complexity cover the whole
spectrum of mathematics. The main application of graph theory in
this field is as a source of examples. Whether this is due to the
apparent simplicity of hard graph problems or to some inherent
property of graphs or is just an historical accident is hard to
say. However it is surprising how many of the key problems in
complexity are best illustrated by graphical examples. In this
expository talk I shall attempt to interest graph theorists in
complexity by discussing some less well-known problems of a graph-
theoretic nature.

2. BASIC CONCEPTS

A precise definition of the amount of time and space used by an
algorithm can be given by implementing the algorithm on a Turing
machine. We will give these precise definitions later but first
proceed informally.

For any property π of graphs which is invariant under isomorphism we define the _time complexity_ $t_n(\pi)$ to be the minimum number
of steps needed by an algorithm to decide whether or not an
n-vertex graph has property π, in the worst case. That is, if
$t(\underline{A},G)$ denotes the time taken by an algorithm $\underline{A}$ which decides
whether or not graph G has property π, then

$$t_n(\pi) = \max_{G} \min t(\underline{A},G),$$

where the minimum is taken over all π deciding algorithms $\underline{A}$ and the maximum is taken over all n-vertex graphs G.

The space complexity $s_n(\pi)$ of π is defined analogously, where $s(\underline{A},G)$ is the amount of work space needed by the algorithm $\underline{A}$. From the practical point of view a problem is tractable if it can be done in polynomial time — that is, if there exists a polynomial p such that, for each integer n, $t_n(\pi) \leq p(n)$. The class of such problems is denoted by P. Non-deterministic polynomial time, NP, can be crudely described as the class of properties which can be checked in polynomial time. Examples are given later.

Probably the most important open problem in theoretical computer science is whether or not P = NP.

As an example, consider the property of having a hamiltonian cycle, a property which we denote by HAMILTONIAN. Then

$$O(n^2) \leq t_n(\text{HAMILTONIAN}) \leq O(n!)$$
$$O(\text{loglogn}) \leq s_n(\text{HAMILTONIAN}) \leq O(n).$$

Thus the property HAMILTONIAN is not known to be in polynomial time (and is indeed conjectured not to be), but is in polynomial space.

For many of the problems of complexity it suffices (certainly at the first stage) to work with loose ideas of space and time, as discussed above. However, when it comes to rigorous details it is usually simplest to go back to a basic model such as the Turing machine. In the next section we discuss more precisely some of the main ideas.

3. COMPUTATIONAL COMPLEXITY — PRECISE DEFINITIONS

We use the Turing machine model of computation to measure time and space complexity. Our Turing machines will have three tapes: a two-way read-only input tape, a one-way write-only output tape, and a two-way read-write work tape. Associated with such a machine are finite alphabets: Σ (input alphabet), Δ (output alphabet), and Γ (work tape alphabet); also a finite set of states, a start state q_0, and a transition function

$$\delta \; : \; Q \times \Sigma \times \Gamma \rightarrow 2^{Q \times \Gamma \times (\Delta \; \{\lambda\} \times \{R,L\}^2)}.$$

Given a state, a symbol being read on the input tape, and a symbol
being read on the work tape, the machine does one of a finite
number of 'moves' each of which consists of going to a new state,
writing a symbol on the work tape, outputting either a symbol or λ,
and moving the input tape and work tape heads. As defined, our
Turing machines are <u>nondeterministic</u>.

A Turing machine is <u>deterministic</u> if the cardinality of
$\delta(q,\sigma,\tau) \leqq 1$ for each triple $(q,\sigma,\tau) \; \epsilon \; Q \times \Sigma \times \Gamma$.

Given an input $x \; \epsilon \; \Sigma \times \Gamma*$, a <u>computation of T on input x</u> is a
finite sequence of configurations of the Turing machine which
begins in the starting configuration (the machine is state q_0, the
input tape contains x with the input head on the first letter of x,
and the other tapes are empty); each other configuration follows
from the previous one via the transition rule, and ends in a configuration from which no configuration can follow.

A Turing machine T <u>runs in time t</u>: $N \rightarrow N$ if, for each n and
each $x \; \epsilon \; \Sigma*$ such that $|x| = n$, every computation of T on input x
has length $\leqq t(n)$.

A Turing machine T <u>runs in space s</u>: $N \rightarrow N$ if, for each n and
each $x \; \epsilon \; \Sigma*$ of length n, at most $s(n)$ distinct tape cells on the
work tape are used in each computation of T on input x.

A <u>set</u> $L \subseteq \Sigma*$ is <u>computable in nondeterministic time</u> (<u>space</u>) <u>r</u>
if there is a Turing machine T that runs in time (space) r such
that for all $x \; \epsilon \; \Sigma*$, $x \; \epsilon \; L$ if and only if there is a computation
of T on input x such that T outputs the symbol for ACCEPT.

A <u>set</u> L <u>is computable in time</u> (<u>space</u>) <u>r</u> if in the above definition the Turing machine is deterministic.

A <u>function f</u> : $\Sigma* \rightarrow \Delta$ <u>is computable in time</u> (<u>space</u>) <u>r</u> if there
is a deterministic Turing machine T that runs in time (space) r
such that, for all $x \; \epsilon \; \Sigma*$, when T halts on input x the machine has
outputted the string $f(x)$.

We define NP = NP-TIME (NP-SPACE) to be the class of sets L
such that there is a polynomial p such that L is computable in
nondeterministic time (space) p. Similarly, P = P-TIME(P-SPACE)

is the class of sets L such that there is a polynomial p such that L is computable in time (space) P.

Theorem 1 (Savitch). P-SPACE = NP-SPACE.

Theorem 2. P $\subseteq$ NP $\subseteq$ P-SPACE.

If s: N $\to$ N, then define (N)SPACE(s(n)) to be the class of sets computable in (nondeterministic) space s.

Given sets L ε Σ* and M ε Δ*, we say that L is <u>log space reducible</u> to M (L $\leq_{log}$ M) if there is a function f: Σ* $\to$ Δ* such that f is computable in space log and, for all x ε Σ*, x ε L if and only if f(x) ε M. The relation $\leq_{log}$ is reflexive and transitive.

Let <u>S</u> be a class of sets. A set L <u>is log space complete in</u> <u>S</u> if L ε <u>S</u> and, for all M ε <u>S</u>, M $\leq_{log}$ L.

Theorem 3 (Cook). There exist complete languages in NP-TIME.

Theorem 4 (Meyer and Stockmeyer). There exist complete languages in NP-SPACE (= P-SPACE).

The importance of completeness is illustrated in the following remarks:

(a) if L is log space complete in NP, then L ε P if and only if
     P = NP;

(b) if L is log space complete in P-SPACE, then
     (i) L ε NP if and only if NP = P-SPACE;
     (ii) L ε P if and only if P = P-SPACE.

Other forms of reducibility exist. Probably the most common is the following:

The language L_1 over Σ is <u>polynomially reducible</u> to L_2 over Δ if there exists f : Σ* $\to$ Δ* such that x ε L_1 if and only if f(x) ε L_2 and f is computable in polynomial time.

Theorem 5. If L_1 is logspace reducible to L_2, then L_1 is polynomially reducible to L_2.

For more details about these basic ideas we refer to the books by Garey and Johnson [3] and Hopcroft and Ullman [4].

4. LOWER BOUNDS

A major problem in complexity is the lack of techniques for getting lower bounds for the space and time complexity of a problem.

As far as time is concerned, there is one crude lower bound based on the following remark: "For most graph and digraph properties π, the time complexity $t_n(\pi)$ satisfies

(1) $$Kn^2 \leq t_n(\pi),$$

where K is some non-negative constant."

Proof. In general, a property π of graphs cannot be decided without every entry in the input being examined for at least one possible input graph G. Since the input is the adjacency matrix of the graph or digraph it has $O(n^2)$ entries and the Turing machine or computer would need to make at least this number of moves in order to read the input data. //

However a word of warning here: some non-trivial properties can be decided without examining all the entries in the adjacency matrix. As an example of this, consider the problem of deciding whether a digraph D has a vertex v (called a sink) such that, for each vertex u of D with u $\neq$ v, there exists a directed edge (u,v) and there does not exist a directed edge (v,u).

The reader will quickly verify that whether or not a digraph has a sink can be decided by examining relatively few (in fact, $3n - \lceil \log n \rceil$) of the entries in the adjacency matrix. A graph property which cannot be decided without examining all the data is called elusive. For a discussion of elusive properties and some interesting open problems, see Bollobás [2, Chapter 8]. For example, there is still no clear understanding of what makes a property of graphs elusive, although there is a fairly natural belief that most graph properties are elusive.

As far as space is concerned there are even fewer techniques available. An important 'gap theorem' due to Hartmanis and Stearns (see [4]) can be loosely described as follows: "If a property cannot be decided in constant space then it needs space at least $O(\log\log n)$."

We should emphasize here that these lower bounds for time and space are based on the assumption that the input format for all our problems is the adjacency matrix of the graph or digraph. If we use what is known as 'edge list' input format, it is possible to get algorithms for non-trivial properties such as planarity which have time complexity $O(n)$, in contrast to (1) above. However, as regards the coarse grid of polynomial or nonpolynomial complexity to be discussed next, such differences of input format do not matter.

5. POLYNOMIAL-TIME AND LOG-SPACE

It is straightforward to show that if a problem π can be decided in space $s(n)$ then there exists a constant C such that

$$t_n(\pi) \leq c^{s(n)}.$$

The idea of the proof is that the right-hand side is a bound on the total number of configurations in which the machine can find itself. Hence any algorithm using more than this amount of time must be in exactly the same configuration twice, and so a shorter algorithm exists.

Consequently we know that L = Logspace is a subset of P = P-time. A long-standing and hard problem is to prove:

CONJECTURE. Log space is a proper subset of Polynomial time.

Indeed only very simple properties of graphs are known to be decidable in log space.

Example: it is easy to decide whether a graph has all its vertex degrees even, using only logspace.

However, at the moment at least, no algorithm which uses only logspace is known for the very simple graph properties of being eulerian, bipartite, or even connected. We pose in particular the questions:

Problem 1. Do there exist algorithms using only logarithmic space for deciding whether a graph G is
(a) a tree, or (b) connected?

80

Problem 2. Show that there is no logspace algorithm for the problem HAMILTONIAN.

For reasons which should become clearer later, settling Problem
2 in either direction would be a major step forward in computational
complexity.

6. NONDETERMINISTIC POLYNOMIAL TIME — NP

The formal definition of the class NP is given in Section 3. More
loosely, a property $\pi \in$ NP if there exists some 'certificate' $\sigma(\pi)$
of membership of π for which, once σ has been guessed, the verification that it is actually a certificate for π can be carried out
in polynomial time. We illustrate by some examples:

Example 1. If π is the property HAMILTONIAN, then a certificate
for π would be any hamiltonian cycle, since the verification that a
given set of edges constitutes a hamiltonian circuit is obviously a
polynomially bounded computation.

Example 2. If π is the property 3-COLOURABLE of being a
3-colourable graph, then a certificate would be any legitimate
3-colouring of the graph.

Example 3. Consider the properties NONHAMILTONIAN and NON-3-
COLOURABLE. In both cases no-one as yet knows of certificates
which would show them to be in the class NP. Indeed, they are
thought not to be in the class NP.

Let co-NP be defined to be the class of properties complementary
to those in NP; a major open question is to prove that NP $\neq$ co-NP.

Problem. It is known that NP and co-NP are subsets of P-SPACE
(polynomial space - see Section 3), and it is conjectured that
both are proper subsets. However in the hierarchy shown by the
following ordering,

$$L \subseteq P \subseteq NP \cap \text{co-NP} \quad \begin{array}{c} NP \\ \subseteq \quad \subseteq \\ \subseteq \quad \subseteq \\ \text{co-NP} \end{array} \quad \text{P-SPACE,}$$

all that is known is that Logspace is a proper subset of P-SPACE.

Any result concerning this hierarchy would be of major importance, and thus our earlier question of whether or not HAMILTONIAN could be done in logspace can be seen as a way of separating the class NP from the class L.

Consider the following problem which we call GRAPH PRODUCT.

Input: Graph G.

Question: Is G the cartesian product of two smaller graphs G_1 and G_2?

It is trivial to see that GRAPH PRODUCT is a member of NP — a certificate is the two factors. However, I see no way of showing that it is in co-NP, and nor do I see how to show that it is NP-complete.

The reason why I am interested in this particular problem is that I know no problem concerning graphs which is in NP ∩ co-NP but which is not known to be in P. The most well-known problem with this property is the problem PRIME which asks whether or not a given integer is prime. The above is (at a very heuristic level) a graphic analogue of this number-theoretic problem. Hence I pose the problem:

Problem. What is the status of GRAPH PRODUCT in the complexity hierarchy?

Another problem which originally arose in connection with joint work with A.J. Mansfield [5] is the following:
Consider the property CRITICAL, defined to be the class of graphs which are critically k-colourable for some integer k. It is straightforward to prove:

Proposition. If CRITICAL ε NP then NP = co-NP.
Since it is thought unlikely that NP = co-NP, this is fairly strong evidence that CRITICAL $\notin$ NP. However I have no similar evidence about co-NP, and therefore pose the following problems:
(A) Is CRITICAL ε co-NP? — that is, is there any certificate of non-criticality which can be checked in polynomial time?

A purely graph-theoretic problem which arose in [5], and which does not seem to have been considered in the literature, is the following:

(B) If c(n) denotes the number of critical graphs on n vertices, what is the order of magnitude of c(n)?

I close this section with a discussion of a problem about hamiltonian graphs, which A.J. Mansfield and I have attacked without success.

If a graph has exactly one hamiltonian cycle we say it belongs to the class UNIQUE HAMILTONIAN.

Problem. Is UNIQUE HAMILTONIAN a member of the class NP or of the class co-NP?

Intuitively, graphs with this property would appear to be so few and so restricted that there might be some hope of characterizing them. This would make it unlikely that deciding whether a graph was uniquely hamiltonian is harder than deciding whether it was hamiltonian. On the other hand it could be argued that 'uniquely hamiltonian', being a more restrictive property, should be even harder to characterize!

7. RANDOMNESS

Probabilistic algorithms have become increasingly important in the last few years. The most well-known of these is the Rabin-Solovay-Strassen randomized algorithm for deciding whether or not an integer is prime; for a discussion of these and other randomized algorithms, see [6]. Here we discuss some attractive graph problems which arise due to a recent randomized algorithm for deciding whether or not a graph is connected.

Theorem (Aleliunas et al. [1]). There is a randomized algorithm which works in polynomial time and logspace for deciding whether a graph is connected.

Related to this problem is the following very attractive idea.

Let G be a regular graph of degree d, and label the edges at each vertex with the integers in {1, 2, ..., d}, each edge being distinctly labelled. Thus each edge has two labels, one at each

end. A sequence σ of symbols from the list {1, ..., d} <u>traverses</u>
G if, starting from any vertex of G and following the edge labels
in σ, we visit each vertex of G. A sequence is <u>n-universal</u> if it
traverses <u>every</u> n-vertex d-regular graph for <u>every</u> possible labelling. Now it is clear that there do exist n-universal sequences,
though these 'obvious' sequences are very long, at least of
exponential length. However in [6] is proved the following
theorem:

<u>Theorem.</u> There is an n-universal sequence of length $O(n^3 \log n)$.
The proof is purely probabilistic, and shows the existence but
gives no insight into how to construct one. This suggests the
following problems:

<u>Problem.</u> Find a constructive proof of the above theorem.

<u>Problem.</u> Find a construction which gives universal sequences of
polynomial length (say, $O(n^{10})$).

<u>Problem.</u> Find an algorithm which works in logspace and which suc-
cessively works out the terms of an n-universal sequence.

A solution to this last problem would be very interesting since
it would give a logspace algorithm (essentially a look-up table)
for deciding whether a graph is connected.

Finally I know no good lower bound for the length of universal
sequences. Must every universal sequence have length at least $\binom{n}{2}$?

REFERENCES

1. R. Aleliunas, R.M. Karp, R.J. Lipton, L. Lovász and C. Rackoff,
        Random walks, traversal sequences and the
        complexity of maze problems, Proc. 20th IEEE
        Symp. on Foundations of Computer Science
        (1979), 218-223.

2. B. Bollobás,   Extremal Graph Theory, Academic Press, New
        York,   1979 .

3. M.R. Garey and D.S. Johnson, Computers and Intractability,
        Freeman,   1979 .

4. J.E. Hopcroft and J.D. Ullman, Introduction to Automata Theory,
        Languages and Computation, Addison-Wesley,
        Reading, Mass.,   1979 .

5. A.J. Mansfield and D.J.A. Welsh, Some colouring problems and
                     their complexity, Proc. Cambridge Conference,
                     1981 (ed. B. Bollobás), North-Holland,
                     Amsterdam, 1982.

6. D.J.A. Welsh,   Randomised Algorithms, Proc. Conf. on
                     Combinatorial Optimisation (Stirling), 1981,
                     to appear.

D.J.A. Welsh
Merton College
Oxford

7 Some chemical applications of the eigenvalues and eigenvectors of certain finite, planar graphs
R.B. MALLION

1. INTRODUCTION

The eigenvalue spectrum of a graph is a graph invariant; the actual
form of the eigenvectors depends, albeit predictably, on the way in
which the vertices of the graph are labelled. However, although
the set of eigenvectors and the family of corresponding eigenvalues
may be easily obtained — even for large graphs — by means of standard computational techniques, the former, in particular, have sel-
dom been very intensively exploited by Mathematicians.

In fact, Chemists have routinely been using the eigenvalues and
eigenvectors of the graphs that represent the carbon-atom connec-
tivities of planar, conjugated hydrocarbon-molecules, ever since
the classic quantum-mechanical work of Hückel, Coulson and others
in the 1930s and 1940s (see, for example, [1]). The eigenvalues
correspond to quantum-mechanically allowed energy-levels within
the associated chemical-species, while the components of the
individual eigenvectors represent the weightings of the contribution
of the so-called 'atomic orbital' centred on each carbon atom of
the hydrocarbon in question to a molecular orbital considered to
extend over the whole framework of the system (see [1-11]). By
admitting the possibility of arbitrarily edge- and vertex-weighted
graphs, we can treat conjugated molecules other than just hydrocarbons [12-15].

We shall outline the formal connection between the methods
developed by these early Quantum Chemists — entirely independently
of Graph Theory — and those employed in abstract Graph Theory (a
connection only fully capitalised upon within the last ten years).
The Hückel method, which forms the basis of much of the discussion

that is to follow, is, however, now somewhat _passé_ in modern
Theoretical Chemistry, despite the amount of attention still given
to it in the chemical literature. This attention is due not to
the current relevance of Hückel Theory as such, but arises precisely
because interest in its graph-theoretical foundation — first mentioned in some pioneering papers in the 1950s and 1960s (see [2-5])
— has only relatively recently been _re_-aroused (see, for example,
[6-12]). Accordingly, I shall propose the possibly provocative
and somewhat controversial thesis that a greater advance in pure
Graph Theory is likely to arise from the way in which Chemists
have independently developed and applied ideas that have only
subsequently been realised to be graph-theoretical in nature, than
will occur in Chemistry as a result of this rather _arrière-pensée_
realisation.

2. THE HÜCKEL METHOD FOR HYDROCARBONS AND ITS RELATION TO GRAPH THEORY

The neutral carbon atom has a chemical valency of four; the hydrogen
atom has a valency of one. In an 'aromatic' hydrocarbon such as
naphthalene, $C_{10}H_8$, all the carbon and hydrogen atoms that con-
stitute the molecule lie in a unique plane (the _molecular plane_).
The quantum mechanics of the problem [1] determines that the carbon
atom uses three of its valence electrons (negatively-charged
particles, a pair of which between two atoms enables a bond to be
formed between those atoms) to form three equivalent bonds _in_ the
molecular plane; they are technically called "sp_2-hybrid sigma-
bonds", and we shall refer to them simply as _sigma bonds_; for
minimum energy, they make 120° angles with each other, as
schematically shown in Figure 1.

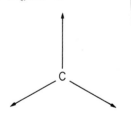

Figure 1

The coplanar sigma bonds that each carbon atom can form in the sorts of molecule under consideration may be to hydrogen atoms and/or to other carbon atoms, as, for example, in naphthalene (Figure 2):

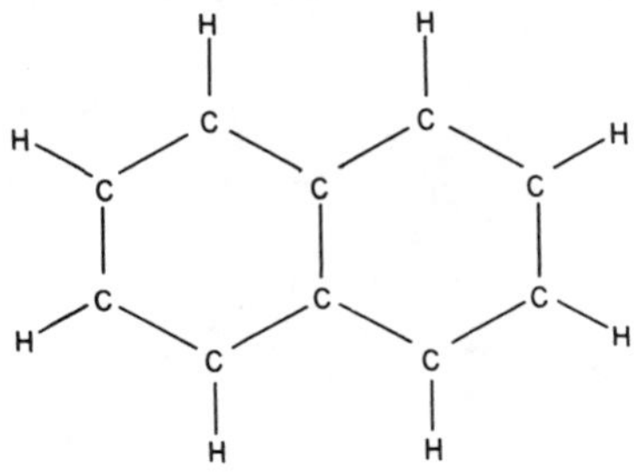

Figure 2

In such a molecule, each carbon atom is bonded by a sigma bond to two or three other carbon atoms, and to at most one hydrogen atom. This leaves one electron, as yet unused and thus still available for bonding, that is possessed by (and centred on) each carbon atom. In the classical 19th-century treatment of Kekulé [1] (and in one of the early quantum-mechanical methods, known as the 'resonance theory' variant of the 'valence-bond' approach [16]), this 'fourth valence' of each carbon atom is accounted for by inserting so-called "double bonds" in the structure; in naphthalene, for example, there are three ways of doing this, as shown in Figure 3.†

---

† The phenomenon of formally <u>alternating</u> 'single' and 'double' bonds, illustrated by the Kekulé structures for naphthalene shown in Figure 3, is referred to by Chemists as 'conjugation'; (geometrically) planar chemical species displaying such conjugation — which form the basis of the whole discussion presented here — are described generically as 'conjugated molecules' or, allowing for the possibility of their being positively or negatively charged, as 'conjugated systems'.

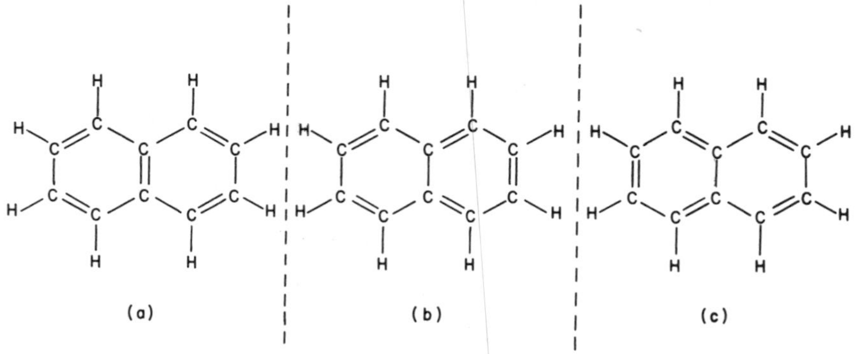

Figure 3

However, in the quantum-mechanical <u>Molecular-Orbital theory</u>, with which we are concerned here, this fourth valence electron is dealt with in an entirely different way. Quantum-mechanically it transpires that, in contradistinction to the electrons forming the carbon-carbon and carbon-hydrogen sigma bonds that lie entirely within the molecular plane, this fourth electron (called, for symmetry reasons, a <u>pi-electron</u>) has an arbitrarily high probability of being found within a 'dumb-bell'-shaped volume, centred on the particular carbon atom in question, and extending <u>above</u> and <u>below</u> the molecular plane. Figure 4 is an attempt to represent this pictorially.

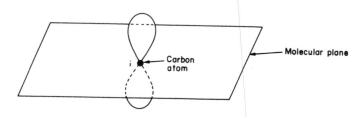

Figure 4

The probability-density function (or <u>wave function</u>) associated with this pi-electron is called the <u>atomic orbital</u> centred on the i-th carbon atom; it is an analytical function of the spherical polar coordinates ($\underline{r}$, θ, ϕ) established with respect to the i-th carbon atom as origin. However, its exact form does not concern us here, and we simply denote such an atomic orbital by the symbol ϕ_i. In a molecule such as benzene, C_6H_6, a perspective view of the six carbon atoms, each bonded to two neighbouring carbon-atoms and a peripheral hydrogen atom and each bearing a carbon pi-electron atomic orbital centred on it, may be imagined to be as in Figure 5.

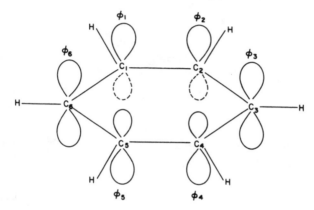

Figure 5

But the above picture does not represent the actual situation that is thought, on the present model, to obtain in benzene; in the quantum-mechanical model of such molecules developed by Hückel [1], it is assumed that these pi-electron atomic-orbitals, each localised on its own particular carbon atom, may combine together to form what is known as a <u>delocalised molecular-orbital</u> that is considered to extend over the entire sigma-bonded carbon-atom framework of the molecule. This means that an electron that was previously confined to, say, the atomic orbital centred on the carbon atom labelled 3 in Figure 5 is now contributed to a pi-electron 'pool' and is able to migrate to the vicinity of other carbon atoms within the ring, thereby reducing the energy of the system and giving the molecule as a whole the peculiar stability that these so-called 'aromatic hydrocarbons' have long been known to have. This would then give rise to two toroidal (doughnut-shaped) regions of electron density, above and below the plane of the molecule, as depicted in Figure 6.

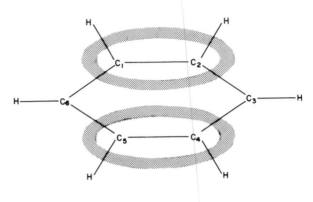

Figure 6

It is seen, therefore, that on this naïve and somewhat simplistic model, the 'mobile' pi-electrons are considered to move over a fixed network of carbon atoms, held together as a rigid framework by the sp_2-hybridised sigma-bonds between the carbon atoms.

Mathematically, if $\{\phi_i\}$ (i = 1, 2, ..., 6) denote the (ortho-normal) <u>atomic</u> orbitals on the six carbon atoms of benzene, we seek an appropriate <u>linear combination</u> of them, to form a suitable <u>molecular</u> orbital, Ψ: that is,

$$\Psi = c_1\phi_1 + c_2\phi_2 + \dots + c_6\phi_6, \tag{1}$$

where the $\{c_i\}$ (i = 1, 2, ..., 6) are 'weighting coefficients' that have, by some criterion, to be found. The criterion used by quantum mechanics is that of the <u>Variation Principle</u> (see [1], [16]), in which the $\{c_i\}$ are varied in such a way that molecular orbitals with quantum-mechanically allowed energies are produced. It turns out that, on this model, there are six such quantum-mechanically allowed energy-levels and that <u>each one corresponds to an eigenvalue</u> <u>of the graph that represents the sigma-bond carbon-atom connectivity</u>

92

of the molecule in question. For example, for benzene (Figure 7(a)), the appropriate graph is that in Figure 7(b).

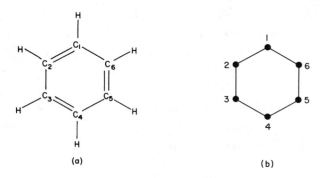

(a)

(b)

Figure 7

There are therefore <u>six</u> different combinations of the $\{\phi_i\}$ (i = 1, 2, ..., 6), each giving rise to a quantum-mechanically allowed molecular-orbital, Ψ_J. We see, therefore, that an extra subscript should have been used in equation (1), the latter now being written:

$$\Psi_J = c_{J1}\phi_1 + c_{J2}\phi_2 + \cdots + c_{J6}\phi_6$$

$$= \sum_{i=1}^{6} c_{Ji}\phi_i. \tag{2}$$

To each eigenvalue, λ_J (J = 1, 2, ..., 6), of the six-vertex graph in Figure 7(b) there therefore corresponds, on the Hückel model, a <u>molecular orbital</u> Ψ_J, the energy of which is related to λ_J, in a way that is explained below. Furthermore, <u>the components of the eigenvector</u>

$$\underset{\sim}{C}_J = (c_{J1}, c_{J2}, \ldots, c_{J6})^T$$

belonging to the J-th eigenvalue λ_J are, on this model, precisely the $\{c_{Ji}\}$ (i = 1, 2, ..., 6) weighting-coefficients that feature in equation (2). This result is perfectly general; if we start with a molecule containing N carbon atoms, there are N atomic orbitals which can be formed, by means of N different linear combinations like the one in equation (2), into N different molecular orbitals, the energies of which are related to the several eigenvalues of the associated molecular-graph.

Thus, in order to perform a Hückel molecular-orbital calculation on a general polycyclic hydrocarbon like naphthalene (Figure 2), composed of N carbon atoms, the required stages are:

1. Form an arbitrarily labelled graph (called the molecular graph) that represents the carbon-carbon sigma-bond skeleton of the hydrocarbon in question; (the hydrogen atoms are suppressed in such a graph). Thus, an appropriate molecular graph for naphthalene is the one depicted in Figure 8:

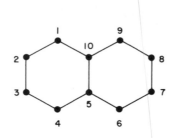

Figure 8

2. Find the eigenvalues λ_J (J = 1, 2, ..., N), and corresponding eigenvectors

$$\underset{\sim}{C}_J = (c_{J1}, \ c_{J2}, \ \ldots, \ c_{JN})^T, \ J = 1, 2, \ldots, N$$

of the molecular graph.

3. The energy ε_J of the J-th molecular orbital Ψ_J is then

$$\varepsilon_J = \alpha + \lambda_J \beta \tag{3}$$

In equation (3), α and β are constants, having the dimensions of
energy. α represents the energy of a pi-electron situated in an
isolated carbon-atom atomic-orbital; the term $\lambda_J \beta$ thus amounts to
the difference in energy between an electron in the molecular
orbital Ψ_J which is delocalised over the whole molecular framework,
and an electron in an isolated atomic-orbital centred on a single
carbon-atom. β is actually a negative constant; hence, the
molecular orbitals of lowest energy are, in fact, associated with
the (algebraically) largest eigenvalues of the molecular graph.

4. The linear combination of atomic orbitals $\{\phi_i\}$ (i = 1, 2, ..., N),
out of which the molecular orbital Ψ_J of energy $\alpha + \lambda_J \beta$ is to be
constructed, is given by a generalisation of equation (2):

$$\Psi_J = \sum_{i=1}^{N} c_{Ji} \phi_i, \tag{4}$$

where the $\{c_{Ji}\}$ (i = 1, 2, ..., N) are the corresponding components
of the eigenvector $\underset{\sim}{C}_J$ (belonging to eigenvalue λ_J) found in stage 2.

The number c_{Ji} (the i-th component of an eigenvector belonging
to the eigenvalue λ_J) may thus be interpreted, on the Hückel
molecular-orbital model, as the weighting of the contribution of
the atomic orbital centred on the carbon atom labelled i in the
corresponding molecular-graph to a molecular orbital Ψ_J of energy
$\alpha + \lambda_J \beta$ that extends globally over the whole framework of the
molecule.

Mention must be made here of how the Hückel model deals with
the fact that not all the λ_J for a particular molecular graph may
be unique — that is, that the spectrum of the graph might contain
multiple eigenvalues. If this occurs, it must mean that two or
more different linear-combinations of the basis atomic-orbitals $\{\phi_i\}$
will give rise to allowed molecular-orbitals with identical
energies. This is perfectly permissible on the model, and there

are no problems about it; Chemists call such molecular orbitals of
the same energy (arising from multiple eigenvalues in the associated
molecular-graph) degenerate molecular-orbitals. There is, however,
the consequence that if this happens, the eigenvectors associated
with such repeated eigenvalues (and hence the linear-combination
weighting-coefficients derived from their components and required
in equation (4)) are not unique. This potential difficulty is
circumvented by our agreeing always to choose such eigenvectors so
that they are mutually orthogonal, both to each other and (as is
automatically guaranteed) to eigenvectors belonging to other eigenvalues. From the well-known properties of real symmetric matrices
(such as the adjacency matrices of undirected graphs), the eigenvectors belonging to distinct eigenvalues are necessarily mutually
orthogonal. If, however, the eigenvalue λ_J occurs m times in the
spectrum of the molecular graph in question, there are m linearly-
independent eigenvectors, $\underset{\sim}{C}_J^{(1)}$, $\underset{\sim}{C}_J^{(2)}$, ..., $\underset{\sim}{C}_J^{(r)}$, ..., $\underset{\sim}{C}_J^{(m)}$, all
giving rise to the same eigenvalue λ_J, but they are not in general
mutually orthogonal. Each of these eigenvectors is, of course,
orthogonal to every eigenvector belonging to all the other eigen-
values, λ_R ($\neq \lambda_J$), but, in general, $\underset{\sim}{C}_J^{(1)}, \underset{\sim}{C}_J^{(2)}, ..., \underset{\sim}{C}_J^{(r)}, ..., \underset{\sim}{C}_J^{(m)}$,
are not orthogonal amongst themselves. However, we can always
choose a new set of eigenvectors, $\underset{\sim}{C}_J'^{(1)}, \underset{\sim}{C}_J'^{(2)}, ..., \underset{\sim}{C}_J'^{(r)}, ..., \underset{\sim}{C}_J'^{(m)}$,
by taking different linear combinations of the $\{\underset{\sim}{C}_J^{(r)}\}$, that are
mutually orthogonal. In all that follows, therefore, when eigen-
vectors are referred to, it will be implicitly assumed that this
process of orthogonalisation has already been carried out; in
addition, it is convenient to agree that all eigenvectors under
discussion have been normalised (that is, scaled so as to be of
unit length). This presumption of the availability of such
orthonormal eigenvectors is vital for the validity of the results
and formulae presented in Section 5.

3. EXTENSION TO EDGE- AND VERTEX-WEIGHTED MOLECULAR-GRAPHS
In the previous section, it was shown how a Hückel molecular-
orbital calculation on a conjugated hydrocarbon reduces simply to
the graph-theoretical problem of finding the eigenvalues and eigen-

vectors of the simple connected unweighted graph that represents
the carbon-atom connectivity of the molecule under consideration.
But there is no reason to confine the treatment to unweighted
molecular-graphs. For example, it is clear that the edges (bonds)
(1, 2), (3, 4), (6, 7) and (8, 9) in the naphthalene molecular-
graph shown in Figure 8 are all symmetrically (and hence chemically)
equivalent, and are in different electronic environments from the
set of edges {(1, 10), (9, 10), (4, 5), (5, 6)}; similarly
{(2, 3), (7, 8)} are a set of equivalent edges, while the edge
(5, 10) is in a symmetry class of its own. There might, therefore,
be a very sound chemical reason for wanting to weight these various
sets of edges differently in the associated molecular-graph; we are
therefore at liberty to do this by choosing the corresponding off-
diagonal elements of the adjacency matrix to be something other
than 0 or 1, according to some physical criterion, if we feel from
subjective chemical considerations that it is appropriate to do so;
in other words, we may arbitrarily weight the edges of the molecular
graph, without changing in any way the interpretations and con-
clusions presented at the end of Section 2.

In the same way, it might be argued that vertices {1, 4, 6, 9}
are symmetrically equivalent (and, therefore, from a chemical
stand-point, are in identical electronic environments), as are the
members of the set {2, 3, 7, 8} and those of the set {5, 10}.
Accordingly, we may wish to weight these sets of vertices differently in the associated molecular-graph. This can be done by
arranging to have the diagonal elements of the adjacency matrix
something other than zero. Again, all of this is perfectly possible
without changing any of the deductions previously made at the end
of Section 2.

In practice, from a chemical point of view, weighting the edges
and vertices of the (bipartite) molecular-graph (Figure 8) that
represents a molecule like naphthalene (Figure 2) might, with some
justification, be considered to be a refinement that the overall
crudity of the Hückel method (viewed as an approximation to a proper
application of the exact Schrödinger equation in quantum mechanics
[1]) does not warrant. However, even on the naive Hückel model,

such edge- and vertex-weightings will certainly need to be invoked
if the method is to be applied to conjugated molecules in which one
or more carbon atoms in the network are replaced by other types of
atoms, such as nitrogen, oxygen, phosphorus, etc.; such atoms are
often referred to as 'hetero-atoms'. For example, consider a
Hückel calculation on pyridene; this molecule is formally obtained
from benzene by replacing one C-H moiety by a nitrogen atom (N).
The structural formula of this molecule is shown in Figure 9(a),
and its associated molecular-graph is given in Figure 9(b). The

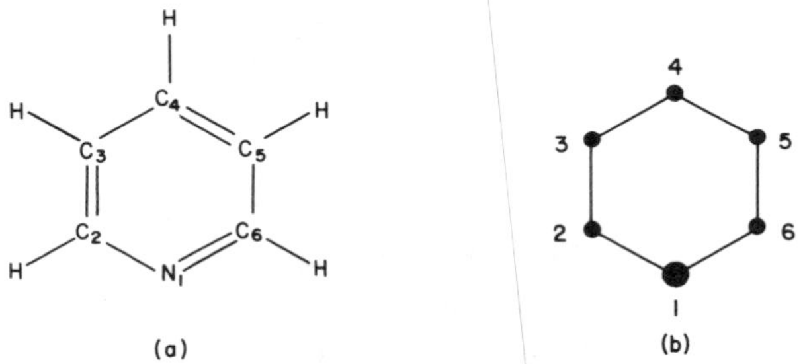

(a) (b)

Figure 9

nitrogen atom, like a carbon atom, contributes one electron to the
mobile pi-electron 'pool' that is able to migrate over the whole
molecular framework, but otherwise, the nitrogen atom has a different attraction for electrons (that is, it has a different
electronegativity [1], [19], [20]) from the carbon atoms in the
pyridene ring. We might, therefore, think it appropriate that the
diagonal element associated with the nitrogen atom in the adjacency
matrix of the corresponding molecular-graph — the (1, 1) element,
on the labelling scheme adopted in Figure 9 — should be different
from the other diagonal elements. Further, the edges {(1, 6),
(1, 2)} in the molecular graph represent carbon-nitrogen sigma-
bonds in the pyridene molecule. Again, therefore, chemical considerations might suggest that different edge-weightings ought to
be invoked for edges representing different types of sigma bond.

98

The actual numerical values to be used for these various edge- and vertex-weightings have to be decided upon from somewhat intuitive chemical arguments, and they are, it must be admitted, largely subjective [1], [19], [20]. However, this need not concern us here; the point that has been made in this section is that by entertaining the notion of arbitrarily edge- and vertex-weighted molecular-graphs [12-15], Hückel molecular-orbital calculations on hetero-conjugated molecules may be carried out just as easily as those on hydrocarbons that were described in Section 2. Moreover, the eigenvalues and eigenvectors obtained from these edge- and vertex-weighted molecular graphs lend themselves to precisely the same physical interpretation as was given (at the end of Section 2) for the eigenvalues and eigenvectors of the unweighted graphs that represent hydrocarbon molecules.

4. THE AUFBAU PROCESS

Now that the allowed energy-levels of the pi-electron system and the linear-combination-of-atomic-orbital (L.C.A.O.) molecular-orbitals associated with them have, in principle, been determined from the appropriate molecular graph, we turn our attention to how a 'pool' of N pi-electrons (one per carbon atom and one per hetero-atom of the sigma-bonded framework) may be assigned to these variationally-determined molecular-orbitals. For this we invoke a concept known in Physics and Chemistry as the Aufbau Principle.

The Aufbau process is a means by which the N available electrons in the pi-electron 'pool' are assigned to the N L.C.A.O. molecular-orbitals $\{\Psi_J\}$ (J = 1, 2, ..., N). This distribution — which yields what is called a 'ground-state electronic configuration' — is conducted according to well-defined quantum-mechanical rules which have a sound basis in Physics and Chemistry. These involve what are known as 'Hund's Rules of Maximum Multiplicity' (see, for example, [1], [20], [23]). We do not justify them here, but merely illustrate the Aufbau process by building up the ground-state electronic-configuration of benzene.

The molecular graph of benzene (Figure 7) has eigenvalue list (2, 1, 1, -1, -1, -2). The energies of the six L.C.A.O. molecular-orbitals are thus, by equation (3), $(\alpha + 2\beta, \alpha + \beta, \alpha + \beta, \alpha - \beta, \alpha - \beta, \alpha - 2\beta)$. (Recall that β is negative, and hence the molecular orbitals in the above list are given in order of <u>increasing</u> energy.) It is conventional to represent these allowed energy-levels as horizontal lines, placed on a vertical energy-scale as in Figure 10. Degenerate molecular-orbitals (arising from multiple eigen-values in the spectrum of the molecular graph — see Section 2) are depicted by means of the appropriate number of horizontal lines at the same height (such as Ψ_2 and Ψ_3, and Ψ_4 and Ψ_5, in Figure 10).

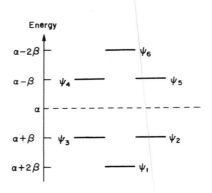

Figure 10

The Aufbau process, for building up the ground-state electronic-configuration of benzene by feeding electrons to these various energy-levels, now begins. There are six pi-electrons to distribute amongst these levels. The first is assigned to the molecular-orbital Ψ_1 of lowest energy $(\alpha + 2\beta)$; so is the second. These two electrons placed in the same molecular orbital must have opposite 'spins', in order to satisfy the relevant quantum-mechanical rules (see [1], [20], [23]). The presence of two elec-trons of opposite spins in the orbital Ψ_1 is conventionally depicted by placing two half-headed arrows, one upright and one upside-down, on the energy level representing the energy of orbital Ψ_1 in Figure 10, as in Figure 11. The orbital Ψ_1 is now said to be 'filled' since, according to the quantum-mechanical rules, no orbital may be assigned more than two electrons. The <u>occupation number</u> ν_1 of the orbital Ψ_1 is 2, because there are two electrons in it.

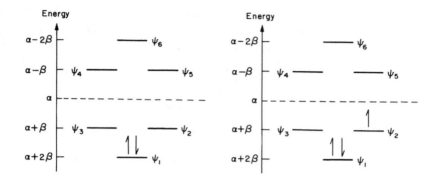

Figure 11   Figure 12

Four electrons from the original six have still to be disposed of. The first of these remaining electrons is assigned to the next-lowest available vacant orbital. This is either of the two 'degenerate' orbitals Ψ_2 and Ψ_3 — let it be Ψ_2; the situation is now as in Figure 12. The next electron goes not into Ψ_2 (where there would be an energy increase due to repulsion with the electron already in that orbital) but into the unoccupied orbital Ψ_3 with the same energy.[†]

At this stage, the electronic configuration is as illustrated in Figure 13. When assignment of the fifth electron is considered, the choice is between placing it in Ψ_2 or Ψ_3, each of relatively low energy ($\alpha + \beta$) — but thereby incurring a repulsion energy

---

[†] This has been called the 'railway-carriage effect', by analogy with the observation that most people, when entering a railway compartment, will sit preferentially in vacant double seats, rather than choose a double seat that is already half-occupied; only when <u>every</u> available seat is at least half-occupied will the passenger reluctantly fill the remaining half of a partially-filled double seat!

between the electron currently being placed and the ones already
occupying these orbitals — and assigning it to sole occupation of
an orbital of much higher energy, Ψ_4 or Ψ_5, each at $\alpha - \beta$. It
turns out that the former alternative is energetically the least
unfavourable, and so the remaining two electrons are assigned,
with paired spins, to Ψ_2 and Ψ_3, as in Figure 14. All six of the
original pi-electrons in the 'pool' have now been assigned, and
Figure 14 thus represents the ground-state, pi-electronic configuration of benzene.

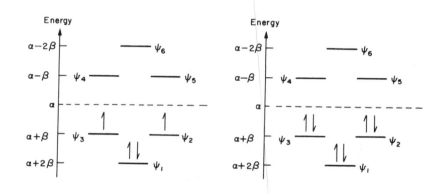

Figure 13   Figure 14

This electronic ground-state is described as $(\Psi_1)^2 \, (\Psi_2)^2 \, (\Psi_3)^2$
$(\Psi_4)^0 \, (\Psi_5)^0 \, (\Psi_6)^0$, where each superscript denotes the number of
electrons assigned by the Aufbau process to the orbital whose
symbol Ψ_J is adorned by it.

The benzene molecule is thus seen to be a stable one because
all the pi-electrons in it are assigned by this process to
orbitals with a lower energy ($\alpha + 2\beta$ and $\alpha + \beta$) than that (α) of a
similar electron in an isolated carbon-atom. The occupation number
ν_J of the orbital Ψ_J may take only the values 0, 1 or 2; for the
ground state of benzene, $\nu_1 = \nu_2 = \nu_3 = 2$; $\nu_4 = \nu_5 = \nu_6 = 0$.

Rouvray and the present author have shown [24], [25] that the
Aufbau process may be simulated by an entirely non-physical

algorithm to yield ν_J-values from an arbitrary eigenvalue-spectrum, as follows:

1. Order and label the eigenvalues of an N-vertex graph so that $\lambda_1 \geq \lambda_2 \geq \ldots \geq \lambda_N$.

2. If N is odd, let a and b be respectively the smallest and largest values of J for which $\lambda_J = \lambda_{\frac{1}{2}(N+1)}$.

3. If N is even and $\lambda_{\frac{1}{2}N} = \lambda_{\frac{1}{2}N+1}$, let a and b be respectively the smallest and largest values of J for which $\lambda_J = \lambda_{\frac{1}{2}N}$.

4. If N is even and $\lambda_{\frac{1}{2}N} > \lambda_{\frac{1}{2}N+1}$, let $a = \frac{1}{2}N + 1$, $b = \frac{1}{2}N$.

5. If $a + b = N + 1$, define $\{\nu_J\}$ (J = 1, 2, 3, ..., N) by the rule that $\nu_J = 2$ if $J < a$, $\nu_J = 1$ if $a \leq J \leq b$, and $\nu_J = 0$ if $J > b$.

6. If $a + b \neq N + 1$, a difficulty arises (see [24-26]), and ν_1, ν_2, ..., ν_N are undefined.

This algorithm shows that, once given the Aufbau process, the $\{\nu_J\}$ may be regarded as entirely graph-theoretical quantities, since they are predetermined by the relative magnitudes and the multiplicities of the eigenvalues of the molecular graph in question. This algorithm also shows that a unique and unambiguous ground-state configuration (that is, a well-defined family of ν_J-values) cannot always be obtained from the spectrum of an arbitrary graph. This aspect has been explored in [24] and [25], and is further developed in [26], but we do not dwell on it here, for it is assumed in this discussion that we are dealing with molecular graphs and these will, by hypothesis, possess uniquely defined ν_J-values since they represent extant molecules which must therefore have unambiguous stable ground-state electronic-configurations.

5. TOTAL PI-ELECTRON ENERGIES, CHARGES, COULSON BOND-ORDERS AND OTHER INDICES OF CHEMICAL INTEREST

Having obtained, for an N-vertex molecular graph, its eigenvalues $\{\lambda_J\}$ (J = 1, 2, ..., N), and corresponding (orthogonalised and normalised) eigenvectors $\{\underset{\sim}{C}_J\}$ as described in Section 2, and its

orbital-occupation numbers $\{v_J\}$ (in Section 4), we are in a
position to appreciate how Chemists apply these graph-theoretical
quantities, by means of the Hückel molecular-orbital model, to
calculate indices of chemical interest. As examples, we consider
total pi-electron energies, charges, so-called Coulson bond-orders,
and we conclude by briefly mentioning some other molecular-orbital
concepts that are numerically attainable from knowledge of these
same graph-theoretical quantities.

(a) Total Pi-Electron Energy

In the first place, we can calculate the total pi-electron energy
of the molecule. Since, on this model, the energy of a pi-electron
may be taken to be the energy of the orbital to which it is assigned
on the Aufbau scheme, this total pi-electron energy may be written

$$E_\pi = \sum_{J=1}^{N} v_J \, \varepsilon_J$$

$$= \text{(from equation (3))} \sum_{J=1}^{N} v_J (\alpha + \lambda_J \beta)$$

$$= \alpha \sum_{J=1}^{N} v_J + \beta \sum_{J=1}^{N} v_J \lambda_J$$

$$= \alpha N + \beta \sum_{J=1}^{N} v_J \lambda_J \tag{5}$$

For benzene, as we have seen, $\lambda_1 = 2$, $\lambda_2 = \lambda_3 = 1$, $\lambda_4 = \lambda_5 = -1$,
$\lambda_6 = -2$; $v_1 = v_2 = v_3 = 2$, $v_4 = v_5 = v_6 = 0$. These yield, from
equation (5),

$$\text{(benzene)} \quad E_\pi = 6\alpha + 8\beta. \tag{6}$$

Now, application of the Hückel method [1] to a single, isolated,
double-bond, as in the molecule ethylene (Figure 15), shows the
pi-electron energy of such a bond to be $2\alpha + 2\beta$.

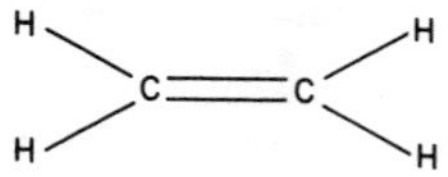

Figure 15

Therefore, if the benzene molecule were to be regarded as three isolated non-interacting double-bonds, as in one of the two Kekulé structures shown in Figure 16, its total pi-electron energy would be $3.(2\alpha + 2\beta) = 6\alpha + 6\beta$.

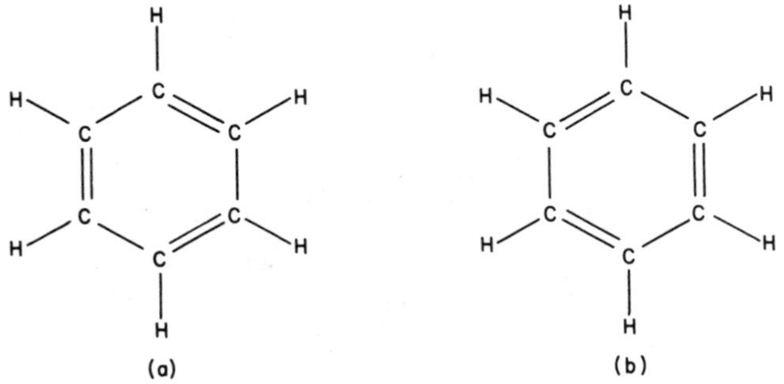

(a)   (b)

Figure 16

The difference between $6\alpha + 6\beta$ and E_π ($= 6\alpha + 8\beta$) is 2β, and can be taken to be a quantitative measure of the extra stability arising from our considering the benzene pi-electrons to be delocalised around the ring, rather than confined within the vicinity of individual 'double' bonds, as in the Kekulé structures shown in Figure 16.

(b) Pi-Electron Charges

The pi-electron charge q_i on the i-th atom of the conjugated system (that is, the i-th vertex of the corresponding molecular-graph), is defined by

$$q_i = \sum_{i=1}^{N} \nu_J \, c_{Ji}^2, \tag{7}$$

where, as explained in Section 2, c_{Ji} is the i-th component of a normalised eigenvector (arranged to be orthogonal to all other eigenvectors) belonging to the J-th eigenvalue. A classic theorem, proved by Coulson and Rushbrooke [27] in 1940, establishes that $q_i = 1$ for all vertices of a (vertex-unweighted) bipartite molecular-graph, whereas for non-bipartite molecular-graphs the q_i-values are, in general, not unity [25]. The q_i-values for an arbitrary (unweighted) non-bipartite graph (taken from [25]) are illustrated in Figure 17. It can be shown [1], [25] that for

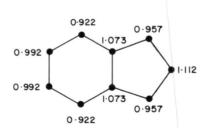

Figure 17

both bipartite and non-bipartite graphs representing neutral molecules,

106

$$\sum_{i=1}^{N} q_i = N, \tag{8}$$

as is to be expected on physical grounds if there is to be 'conservation of charge'. This may be verified in the case of the graph depicted in Figure 17 by simple addition of the nine charges shown.

(c) Coulson Bond-Orders

An index applicable to the sigma bonds of the conjugated system (that is, the _edges_ of the corresponding molecular-graph) is the _Coulson bond-order_ p_{ij} between two atoms (vertices) i and j participating in the conjugation. This is defined by

$$p_{ij} = \sum_{J=1}^{\underline{N}} \nu_J \, c_{Ji} \, c_{Jj}. \tag{9}$$

This quantity has been connected empirically with _bond length_ [1] — the higher the bond-order, the smaller the bond length. There is therefore an important point of philosophy to note here: although we started out with just a graph-theoretical entity (the adjacency matrix of the sigma-bond skeleton of the atoms comprising the conjugated network), we have ended up by being able to make statements — albeit semi-empirical ones — about _geometry_.

(d) Other Molecular-Orbital Indices

There are various other indices calculable, by means of the Hückel molecular-orbital method, from graph eigenvalues, eigenvectors and occupation numbers. These will merely be mentioned here — for the reader will by now have sampled the 'flavour' of these applications, and will be able, if his appetite has been sufficiently whetted, to follow them up in the references given. There are reactivity and free-valence indices, real bond-bond polarisabilities ([1], [7-10], [19], [20]) and, for the calculation of the properties of conjugated molecules in the presence of external magnetic-fields, _imaginary_ bond-bond polarisabilities ([22], [18]).

6. CONCLUSIONS

I have tried to show in this discussion that Chemists have been directly exploiting eigenvalues and, in particular, eigenvectors, of graphs for almost exactly 50 years — in fact, since just before Graph Theory started (with König's book, in 1936) as a modern, formal, mathematical discipline. Indeed, in the course of these 50 years, results have sometimes been discovered by Chemists, in a physical context, that have only later been independently arrived at by Mathematicians as such. The classic example of this is what Chemists call the Coulson-Rushbrooke Theorem [27]. This is a three-part theorem, the first two parts of which are entirely graph-theoretical in nature. The first part states that the eigenvalues of a (vertex-unweighted) bipartite graph occur in pairs, around zero, so that if k is an eigenvalue of the graph in question, $-k$ will also be an eigenvalue; the second part provides a simple relation between the eigenvectors belonging to each such pair of 'complementary' eigenvalues; (several proofs of these two parts are given in [1] and [27]). These two parts of the Coulson-Rushbrooke Theorem were only subsequently (though independently) presented in the Graph-Theoretical Literature — and on several occasions (for example, [28]-[33]; see also [1], [6], [7], [8], [25], [34] and [35] for a history of this theorem). These sections of the theorem have recently been extended to certain vertex-weighted, bipartite graphs [36]. The third part of the Coulson-Rushbrooke Theorem has already been stated in this discussion — it is that the pi-electron charge (as defined in Section 5) on each carbon atom of a neutral, conjugated hydrocarbon whose corresponding molecular-graph is bipartite (what Chemists call an alternant hydrocarbon) is precisely unity. This part of the theorem was not, of course, later independently discovered by Mathematicians, for the latter, unlike Chemists, would not have had the physical motivation to define a concept like atomic charge and then go on to investigate its properties.

It is therefore my contention — and the essence of the thesis being advanced here — that what Chemists have done in the context of Hückel Theory over the last half century, motivated, as they have been, by physical considerations, is likely to be highly material to Graph Theory. Rouvray and I have already proposed [24], [25] that the concept of atomic charge, as defined in Section 5, might usefully be regarded as an entirely abstract, non-physical, graph-theoretical index, and I venture to suggest here that Graph Theorists might do well to examine closely the other molecular-orbital indices (such as bond orders, polarisabilities, etc., referred to in Section 5) that Chemists have devised in the course of their study of planar, conjugated systems.

As I have pointed out elsewhere [37], showing that the simple molecular-orbital theories that Chemists have long used have a graph-theoretical basis has become something of what one might not unfairly call a 'bandwaggon', during the last decade, in the Chemical Literature. However, just as many Graph Theorists do (I feel) fondly imagine that Chemists are far more interested in the counting of hydrocarbon isomers than, in this day and age, they really are, so a considerable number of graph-theoretically inclined Chemists, in their zeal, overestimate the enthusiasm with which fellow Chemists, not so inclined, greet the information that what they have been doing for the last 50 years is essentially topo-logical and graph-theoretical in nature. Indeed, there is a real danger here (because we are engaging in interdisciplinary activity) that Graph Theory — albeit interesting Graph Theory — may mis-leadingly be presented to Chemists as relevant Mathematics, and old-fashioned Chemistry purveyed to Mathematicians as a vital and indispensible application of graph-theoretical ideas.[†] Suspicions of the former have given Graph Theory something of a bad name in the Chemical Literature (Graph Theory being completely synonomous in the minds of many with this somewhat simplistic, intuitive, and

† The actual <u>necessity</u> of formal Graph Theory in the context of simple molecular-orbital theory is perhaps well summed up in a comment from the late Professor C.A. Coulson, F.R.S., who developed and widely applied Hückel Theory in the 1930s and 1940s and was largely responsible for its popularisation within Chemistry: in a letter to the present author dated 1 September 1972 he wrote, concerning Graph Theory: "It's not a field that I know at all, other than as someone who has occasionally had to invent something that I needed for my quantum theory."

now-superceded quantum-mechanical method (Hückel Theory), while I
have been at pains in this discussion to avoid the latter by
emphasising from the beginning that Hückel Theory is largely of
historical (though still, it must be said, very much of pedagogical)
interest in modern Theoretical Chemistry. (An example of a genuine
current chemical application of Graph Theory, not so 'tainted' with
Hückel associations, is the counting of spanning trees, recently
studied by the present author and others [38], [39], [40], [22].)
Therefore, far from extolling the alleged virtues of exploiting
Graph Theory in this particular area of Chemistry, my prime crusade
in this lecture has been to reverse the conventional wisdom and
argue that the way in which Chemists have used graph-theoretical
techniques (albeit largely unknowingly) in the development and
application of Hückel molecular-orbital theory is likely to enrich
pure Graph Theory more substantially than Chemistry will benefit
from the realisation, a posteriori, that these methods independently
devised by Chemists are purely graph-theoretical in nature. In
fact, I would suggest that the modern Graph Theorist specialising
in the spectral theory of graphs might find himself, on encounter-
ing Hückel molecular-orbital theory, in a similar state to that
which Sylvester was moved to describe, over 100 years ago [41],
[42], when he first realised that a chemical structural formula
is essentially a graph: "... I feel as Aladdin might have done in
walking in the garden where every tree was laiden with precious
stones ... There is an untold treasure of hoarded algebraical
wealth potentially contained in the results achieved by the patient
and long-continued labour of our unconscious and unsuspected
chemical fellow-workers."

REFERENCES

1. C.A. Coulson, B. O'Leary and R.B. Mallion, Hückel Theory for
              Organic Chemists, Academic Press, London, 1978.
              (Seventeen major papers on Hückel Theory from
              the period 1931-1950 are listed on page 176 of
              this reference.)

2. K. Ruedenberg, Free-electron network-model for conjugated
systems, V. Energies and electron distributions
in the F.E. M.O. model and in the L.C.A.O. M.O.
model, Journal of Chemical Physics 22 (1954),
1878-1894.

3. H.H. Günthard and H. Primas, Zusammenhang von Graphentheorie
und MO-Theorie von Molekeln mit Systemen
konjugierter Bindungen, Helvitica Chimica
Acta 39 (1956), 1645-1653.

4. K. Ruedenberg, Quantum mechanics of mobile electrons in conjugated bond systems, III. Topological matrix
as a generatrix of bond orders, Journal of
Chemical Physics 34 (1961), 1884-1891.

5. H.H. Schmidtcke, L.C.A.O. description of symmetric molecules
by unified theory of finite graphs, Journal
of Chemical Physics 45 (1966), 3920-3928.

6. A. Graovac, I. Gutman, N. Trinajstić and T. Zivković, Graph
theory and molecular orbitals. Application
of Sachs' theorem, Theoretica Chimica Acta
(Berlin) 26 (1972), 67-68.

7. I. Gutman and N. Trinajstić, Graph theory and molecular
orbitals, Fortschritte der chemischen
Forschung (Topics in Current Chemistry) 42
(1973), 49-93.

8. D.H. Rouvray, The topological matrix in quantum chemistry,
Chemical Applications of Graph Theory (ed.
A.T. Balaban), Academic Press, London, 1976,
175-221.

9. A. Graovac, I. Gutman and N. Trinajstić, Topological Approach
to the Chemistry of Conjugated Molecules,
Springer-Verlag, Berlin, 1977.

10. N. Trinajstić Hückel theory and topology, Semi-Empirical
Methods of Electronic Structure Calculations
— Part A: Techniques, Modern Theoretical
Chemistry, Vol. 7 (ed. G.J. Segal), Plenum
Press, New York, 1977, 1-27.

11. D. Cvetković, M. Doob and H. Sachs, Spectra of Graphs — Theory
and Applications, Deutscher Verlag der
Wissenschaften, East Berlin, 1979, and Academic
Press, London, 1980, 228-244.

12. N. Trinajstić, Computing the characteristic polynomial of a
conjugated system using the Sachs theorem,
Croatica Chemica Acta 49 (1977), 593-633.

13. A. Graovac, O.E. Polansky, N. Trinajstić and N. Tyutyulkov,
Graph theory in chemistry, II. Graph-
theoretical description of heteroconjugated
molecules, Zeitschrift für Naturforschung 29a
(1975), 1696-1699.

14. J.-I. Aihara,   General rules for constructing Hückel
                      molecular-orbital characteristic-polynomials,
                      Journal of the American Chemical Society 98
                      (1976), 6840-6844.

15. M.J. Rigby, R.B. Mallion and A.C. Day, Comment on a graph-
                      theoretical description of heteroconjugated
                      molecules, Chemical Physics Letters 51 (1977),
                      178-182; see also Addendum in Chemical Physics
                      Letters 53 (1978), 418.

16. L. Pauling,   The Nature of the Chemical Bond, 3rd edition,
                      Cornell University Press, Ithaca, New York,
                      1960.

17. C.W. Haigh and R.B. Mallion, Ring-current theories in nuclear
                      magnetic resonance, Progress in Nuclear Mag-
                      netic Resonance Spectroscopy, Vol. 13 (ed.
                      J.W. Emsley, J. Feeney and L.H. Sutcliffe),
                      Pergamon Press, Oxford, 1979, 303-344.

18. R.B. Mallion,   Empirical Appraisal and Graph-Theoretical
                      Aspects of Simple Theories of the 'Ring-
                      Current' Effect in Conjugated Systems, D. Phil.
                      Thesis, Oxford University, 1979.

19. A. Streitwieser, Molecular Orbital Theory for Organic Chemists,
                      John Wiley, New York, 1961 (especially page
                      127).

20. L. Salem,   The Molecular Orbital Theory of Conjugated
                      Systems, W.A. Benjamin, New York, 1966.

21. R.B. Mallion,   Approximate pi-electron 'ring-current' inten-
                      sities in some sulphur heterocyclic analogues
                      of fluoranthene, Journal of the Chemical
                      Society Perkin Transactions II, 1973, 235-237.

22. R.B. Mallion,   Some graph-theoretical aspects of simple ring-
                      current calculations on conjugated systems,
                      Proceedings of the Royal Society of London
                      (A) 341 (1975), 429-449.

23. A. Liberles,   Introduction to Molecular-Orbital Theory,
                      Holt, Rinehart and Winston, New York, 1966,
                      59-66.

24. R.B. Mallion and D.H. Rouvray, Molecular topology and the
                      Aufbau principle, Molecular Physics 36 (1978),
                      125-128.

25. R.B. Mallion and D.H. Rouvray, On a new index for characterising
                      the vertices of certain non-bipartite graphs,
                      Studia Scientarum et Mathematicarum Hungarica
                      13 (1978), 229-243.

26. R.B. Mallion,   An analytical illustration of the relevance of
                      molecular topology to the Aufbau process   (in
                      preparation).

27. C.A. Coulson and G.S. Rushbrooke, Note on the method of molecular orbitals, Proceedings of the Cambridge Philosophical Society 36 (1940), 193-199.

28. L. Collatz and U. Sinogowitz, Spektren endlicher Graphen, Abh. Math. Sem. Univ. Hamburg, 21 (1957), 63-77. (It should be pointed out that some of the work reported in this paper was in fact done in the early 1940s, since Ulrich Sinogowitz was killed in an air raid on Darmstadt on 12 September 1944.)

29. H. Sachs, Über selbstkomplementäre Graphen, Publ. Math. Debrecen 9 (1962), 270-288.

30. A.J. Hoffman, On the polynomial of a graph, American Mathematical Monthly 70 (1963), 30-36.

31. R.B. Marimont, System connectivity and matrix properties, Bulletin of Mathematical Biophysics 31 (1969), 255-274.

32. D. Cvetković, Bihromatičnost i spektar grafa, Mat. Biblio. 41 (1969), 193-194.

33. D.H. Rouvray, Les valeurs propres des molécules qui possèdent un graphe bipartit, Comptes Rendus des Séances de l'Académie des Sciences (Paris) Série C 274 (1972), 1561-1563.

34. R.B. Mallion, Eureka ?, Chemistry in Britain 9 (1973), 242.

35. R.B. Mallion and D.H. Rouvray, The Coulson-Rushbrooke pairing theorem: a case study for a multidisciplinary approach to certain aspects of mathematics and chemistry (in preparation).

36. M.J. Rigby and R.B. Mallion, On the eigenvalues and eigenvectors of certain finite, vertex-weighted, bipartite graphs, J. Combinatorial Theory (B) 27 (1979), 122-129.

37. R.B. Mallion, Review of "Chemical Applications of Graph Theory" (ed. A.T. Balaban), Academic Press, London, 1976, J. Chem. Soc. Faraday Transactions II (1978), 701-702.

38. R.B. Mallion, On the number of spanning trees in a molecular graph, Chemical Physics Letters 36 (1975), 170-174.

39. D.A. Waller, Regular eigenvalues of graphs and enumeration of spanning trees, Atti dei Convegni Lincei 17, Proc. of the Colloq. Internaz. sulle Teorie Combinatorie, Rome, 1973, Accad. Naz. dei Lincei, Tomo I, pp. 313-320.

40. I. Gutman, R.B. Mallion and J.W. Essam, Counting the spanning trees of a molecular graph (in preparation).

41. R.B. Mallion, Euler's Theorem, polyhedra and the phase rule, Chemistry in Britain 8 (1972), 446.

42. J.J. Sylvester, On an application of the new atomic theory to the graphical representation of the invariants and covariants of binary quantics, — with three appendices, Amer. J. Math. 1 (1878) 64-125 = Math. Papers, Vol. 3, 148-206.

R.B. Mallion
The King's School
Canterbury